Communication Skills for the BIOSCIENCES

A graduate guide

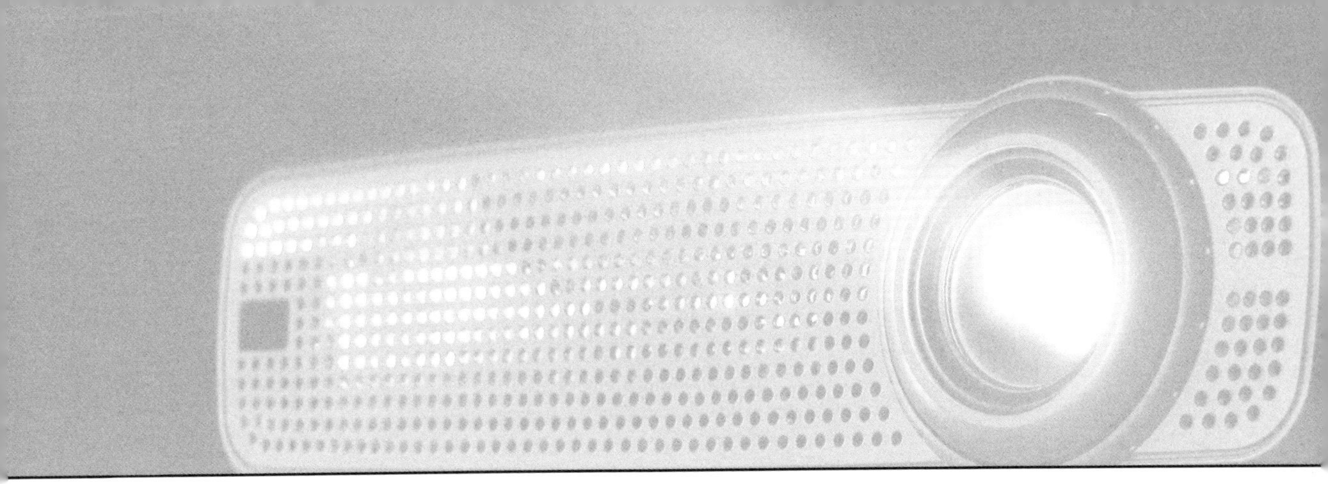

Communication Skills for the BIOSCIENCES

A graduate guide

Aysha Divan

OXFORD
UNIVERSITY PRESS

OXFORD
UNIVERSITY PRESS

Great Clarendon Street, Oxford OX2 6DP

Oxford University Press is a department of the University of Oxford. It furthers the University's objective of excellence in research, scholarship, and education by publishing worldwide in

Oxford New York

Auckland Cape Town Dar es Salaam Hong Kong Karachi Kuala Lumpur Madrid Melbourne Mexico City Nairobi New Delhi Shanghai Taipei Toronto

With offices in

Argentina Austria Brazil Chile Czech Republic France Greece Guatemala Hungary Italy Japan Poland Portugal Singapore South Korea Switzerland Thailand Turkey Ukraine Vietnam

Oxford is a registered trade mark of Oxford University Press in the UK and in certain other countries

Published in the United States
by Oxford University Press Inc., New York

First published 2009

British Library Cataloguing in Publication Data
Data available

Library of Congress Cataloging in Publication Data
Data available

Typeset by Macmillan Publishing Solutions
Printed in Great Britain
on acid-free paper by Ashford Colour Press, Gosport, Hampshire

ISBN 978–0–19–922635–1

1 3 5 7 9 10 8 6 4 2

Contents

Preface

Many graduate and undergraduate programmes provide explicit training in effective communication skills. To be able to communicate effectively you must be fully aware of the different ways in which scientists disseminate data and ideas, and the style associated with each particular type of communication. Scientists communicate primarily through three different media. These are **written** (writing a research paper or a literature review), **oral** (conference presentations, for example), and **visual** (such as the poster presentation). These communications may be through formal channels including scientific meetings and peer-reviewed journals, or through informal channels such as discussion between research group members, your supervisors, and other scientists. The audience with which you communicate may be experts who are working in the same field, scientists who are working in a related field, or a non-specialist member of the public. Therefore, as a scientist, you must be able to communicate your information in an appropriate style taking into account the intended audience.

Who is this book for?

The aim of this textbook is to provide students in the biosciences with the necessary tools to enable them to communicate clearly using written, oral, and visual media, principally to a scientific audience. This book is intended primarily for taught postgraduate and research students. However, it is also useful for undergraduates, particularly at level 3 where more advanced training is required.

What is this book about and how should you use it?

The book is structured thematically into three parts. The first part (Chapters 1–6) covers topics which underpin all forms of scientific communication. These include ethical communication, an overview of the different types of scientific literature, searching and retrieving literature, and critical review of scientific information. The second part of the book (Chapters 7–12) focuses on written forms of communication and includes writing literature reviews, research proposals, research papers, abstracts, and a Masters or PhD thesis. This section also covers how to present your research data in figure or table format. The third part of the book (Chapters 13–15) includes oral and poster presentations and networking.

Throughout the book, there are annotated examples of work selected to illustrate key content and structure of different communication types. These represent examples of good practice and are taken from across the broad range of subject areas that make up the biosciences including biology, biochemistry, biomedical sciences, and microbiology. If you are unfamiliar with the subject matter of any particular example presented in this textbook, then you should still be able to use the descriptive annotations associated with each piece of work to write and organize your own work. However, keep in mind that you may have to adapt the content and structure presented in the examples to fit the specific instructions you have been supplied with. In addition, *checklists* have been included where appropriate to help you self-review drafts of your work. This is an important resource which should assist you to improve the quality of your communication.

Online Resource Centre

www.oxfordtextbooks.co.uk/orc/divan
The Online Resource Centre which accompanies this textbook includes:

For the Student

- Additional examples of materials featured in the book, including a research presentation in PowerPoint (with extra audio commentary)
- Downloadable copies of the *checklists* included in the textbook
- A number of review activities to test your understanding of some of the material covered in this textbook.

You can access these resources at www.oxfordtextbooks.co.uk/orc/divan
You will see the following icon at the point in the book where online materials are available to accompany the text:

For the lecturer

There are also a number of additional online resources for lecturers who have adopted the book for their teaching. These include:

- Figures from the book in electronic format, ready to download
- PowerPoint presentations addressing several of the topics covered in the textbook, such as 'Reviewing scientific literature'.

Why is it important to develop effective communication skills?

Developing strong communication skills will help you achieve the academic requirements of your programme of study. At the same time, the training you receive will build essential skills which are necessary for your career as a scientist. So how might scientific communication benefit you?

It helps to build your reputation as a scientist

Reporting your research findings is the key way by which you build your reputation as a scientist. This can be through oral and poster presentations at scientific meetings or written papers that are published in pee-reviewed journals. Communicating will make you and your work more widely known and improve your chances of securing employment and research funding.

It keeps you informed of the most current developments in your field and related fields of work

An important part of your academic career is to keep up to date with the information in your subject area. Material published in peer-reviewed journals and presented at conferences and seminars represents important and current sources of information. You should therefore read widely and regularly, and take the opportunity to attend as many conferences and seminars as you are able to in order to keep abreast of new findings.

It helps you to build meaningful networks and collaborations with other scientists

Networking involves developing and maintaining contact with people who work in the same or a related field as you. Attendance at conferences, seminars, or workshops can provide an ideal opportunity to meet with and engage in discussion with other scientists in a face to face situation. An additional way of networking is through the electronic medium, using networking sites and discussion groups. The advantage of networking is that it expands the number of people with whom you can share information and ideas. This in turn raises your profile and makes you and your work known to more people within the academic community. The exchanges can be very useful in terms of enhancing your understanding of your work and the work of others. It can also lead to a new job or a meaningful collaboration with other researchers.

You can inform the non-scientific community about developments in science

Contributing to the public understanding of science is an important role of scientists. This can include communicating your work to school children to inform them about developments in science and motivate them to enter a career in science. It could also include communicating your work to adults who are largely unfamiliar with scientific developments and terminology. This in turn could, for example, allow them to make more informed judgements about the relevance and impact of topical issues reported in the media.

Acknowledgements

I wish to thank David Pilbeam, Janice Royds, Mike McPherson, and my four reviewers for their constructive comments—many of their suggestions have been included in the textbook.

I also wish to thank the following people for generously supplying me with examples of work for inclusion in this textbook (these have been used either in the original format or adapted); Alison Baker (Figure 2.1: experimental notebook pages), Brendan Davies (Section 3.7: plagiarism examples), Simon White and Ian Chopra (Section 7.3: literature review), John Grahame (and Craig Wilding) (Section 8.5: grant proposal) Gareth Howell (and the bioimaging and flow cytometry facilities at the University of Leeds) (Figure 11.5), Claire MacDonald (Chapter 13: (online resource) oral presentation), Sam Mason, Jitender Dubey, Rupert Quinnell, and Judith Smith (Figure 14.3: research poster), Brian Wilson (and the Centre for Bioscience ImageBank) (Figure 14.3: photograph of a domestic cat), and Gert Nieuwhof (Figure 14.3: photograph of a sheep). The material on copyright (Chapter 3 (3.6) and Chapter 15 (15.5.2)) has been reviewed (with clarifications and additions) by Eloise Corcoran and Louise Handley (Lee & Priestly, LLP)—many thanks for their contributions.

Finally—my thanks to Jonathan Crowe, OUP publishing editor for providing me with the opportunity to write the book and for his encouraging comments throughout.

Chapter 1

Essential communication skills

Introduction

This chapter introduces the essential skills necessary for effective communication and which underpin the material covered in the chapters that follow.

Specifically, this chapter will:

- describe the basic features associated with scientific communication: accuracy, conciseness, clarity, and source referencing
- provide guidance on selecting and writing references
- describe the three stages a piece of work passes through before it is ready for submission or presentation: planning, drafting, and revising
- provide some guidance on communicating effectively with your supervisor and research group members.

1.1 The essential features of scientific communication

The essential features underpinning all forms of scientific communication are that the work should be reported accurately and concisely, and that any source material used in compiling the work is acknowledged appropriately.

1.1.1 Reporting your work accurately

You must make sure that your work is free from mistakes and from ambiguous comments. It is unlikely that you will intentionally distort facts and figures. However, it is possible for mistakes to occur unintentionally through carelessness, lack of attention to detail, or if you rush your work. If the work is for assessment purposes, then the consequence will be that you lose marks. If the work is published, then the consequence could be that you are accused of deliberately misrepresenting your results or other people may waste valuable time attempting to validate your findings. Either way your reputation as a scientist will suffer. To ensure that your work is accurate:

- Use the international system of units (Système International d'Unités, or **SI units**) when expressing units of measurement and use them correctly and consistently.
- Use abbreviations correctly and consistently.
- Use standard scientific nomenclature when using names of chemicals, genes, chromosomes, and taxonomy.
- Give yourself sufficient time to check drafts of your work for spelling mistakes and grammatical and punctuation errors, and to check the accuracy of your references.
- Check the content of your statements to see that there is no reason for others to misunderstand your work.

If you are unfamiliar with the standard formats listed above, then consult the manual *Scientific Style and Format,* published by the Council of Science Editors (CSE 2006) which describes the standard systems and how to use them correctly.

1.1.2 Reporting your work concisely and clearly

When communicating, you should keep your work concise. This means you should not include superfluous terms and sentences that do not add to the understanding of your work. Instead, use short sentences and avoid unnecessary repetition. Try to use simple words as far as possible and avoid needless technical jargon. There are times, of course, when technical words are necessary, but on the whole avoid long and complex terms.

1.1.3 Referencing source materials in your work

During your academic work you will make use of published material (and sometimes unpublished work) in a variety of ways. For example, you may be asked to review an article critically, write a literature review, propose a programme of research, or present your results. Each of these activities requires that you use the work of others to place **your** work in the context of the wider published literature and to support **your** own analysis, proposals, and interpretations. When using the work of others, you must credit the original author(s) by referencing the source in your work. There are different formats for referencing and these are described in section 1.2. If you do not acknowledge the work of others in your work appropriately, then you could be accused of plagiarism. Advice on how to avoid plagiarism is provided in Chapter 3, section 3.4.

1.1.4 Use of the active voice

Traditionally, scientific articles were written in a passive voice. However, it is becoming much more common to communicate science using an active style, and in fact this is preferred by many journals. For example, it is acceptable to say 'we extracted protein from ...' instead of 'protein was extracted from ....' and 'we conclude ...' instead of 'it was concluded ...' Using active sentences makes communication much more interesting and easier to read. When using an active style of writing, you can use the pronoun 'we' or the pronoun 'I' depending on the purpose for which you are writing. For example, if you are writing a research paper, use 'we' to mean a group of authors. If you are writing a dissertation or thesis, you may use 'we' or 'I'. The 'we' in this context will mean the single author (that is, you and not a group of people). In the case of the dissertation or thesis, the choice between the two pronouns may depend on the preference of your university or department, and you should seek advice from your tutor or supervisor on this before you start to write. It is possible that there are individual journals, departments, or institutions that recommend to their students to report their work using a passive style of writing. In such cases, you must, of course, follow the specific instructions supplied to you.

1.2 Referencing

There are three main reasons for referencing the sources used in your work. One is to credit the original author whose work you are using, the second is to help others identify and locate the source of the information, and the third is to differentiate clearly which part of the work is yours and therefore new, and which is the contribution of others.

There are two parts to referencing completely:

1. Citing the source within the text at the point you refer to it.
2. Listing fully all the sources that you have cited in the text at the end of your writing in the form of a reference list.

1.2.1 Referencing methods

There are several different methods that can be used for referencing, but the two most widely used are the author–date system (also known as the Harvard system) and the numerical system (also known as the Vancouver system).

In the author–date system, the author's surname and the date of publication are given at the relevant point in the text, in parentheses, for each reference cited. All the citations in the text are then placed in a reference list at the end of the work which is organized in alphabetical order by first author's surname. The rules for in-text citation are described in Table 1.1.

In the numerical system of referencing, each reference is cited via a superscript or a bracketed reference number which is inserted at the appropriate point in the body of the text. The reference list is then placed at the end of the work and each citation is listed numerically. The rules for in-text citation are described in Table 1.1.

TABLE 1.1 In-text citation rules for the author–date system and the numerical system of referencing

Author–date system	Numerical system
• If you are citing a source with a single author this can be cited in the text as (Jones, 2005) with the author's surname separated from the year of publication by a comma. If you have already named the author in the text, then only the year needs to be added in parentheses, for example *as discussed by Jones (2005).* • If you are citing a source where there are two authors then both names should be given in the text, for example (Smith and Kent, 2006). • If you are citing a source where there are more than two authors, then the term 'et al.' follows the surname of the first author, for example (Wang *et al.,* 2007). Et al. (an abbreviation of 'et alia') is a Latin term meaning 'and others' and is commonly italicized. • If you are citing two or more different references for the same year and by the same author(s), then each reference should be distinguished from the other by adding a lower case letter after the year. For example the lower case letter 'a' is added to the first 2003 Charnley citation as follows (Charnley, 2003a) and the letter b is added to the second 2003 Charnley citation as follows (Charnley, 2003b). • If you are citing two or more different references which report work on a similar subject matter—the sources can be cited consecutively in the following way (Tsien *et al.,* 1998; Cormack, 1996). You will note that each reference is separated by a semi-colon and ordered in descending chronological order. Ordering in ascending chronological order is also acceptable, but you must always use one style consistently throughout your work. • When citing internet sources in the text cite the author's surname and date in parentheses (if known) or if the personal author is not known then the organization, for example (WHO, 2007).	• The first source cited in your work is allocated number 'one', the second is allocated number 'two' and so on. • If you refer to a source again that has already been allocated a number earlier, then the same number for that particular source is used again. • If you refer to multiple references at the same time, these can be cited together as follows (2,3,4) or (2–4) if they are consecutive. If not consecutive, then they can be cited as (2–4, 14).

1.2.2 Use of reference management software

When compiling reference lists, use reference management software such as EndNote, Reference Manager, or ProCite. This software allows you to:

- Import references from electronic databases directly into a bibliography.
- Organize your references into folders and search and sort through them.

- Insert references into Word documents.
- Format reference lists in different publishing styles.

You are strongly recommended to use reference management software to compile your reference lists. It will make the process less time-consuming and you are less likely to introduce errors into the reference. If you are unfamiliar with how to use a reference software package, then your university may provide training sessions or you can browse the manufacturer's website which will contain information on how to use their software.

1.2.3 Writing references

If you compare reference lists from different sources you will see that there are slight variations in the way in which each reference is written. Some may include the title of articles but others may not (the latter is the style used by the journal *Science*). Some may position the year of publication at the end of the reference, while others may position it after the authors names. Some journals use abbreviated forms of journal titles, while others will use the full journal name. Despite these slight variations in order of arrangement and the amount of information included, all references contain the following essential information: **the names of authors and when and where it was published.** This information is necessary to help you locate the reference. Table 1.2 demonstrates how full descriptions of references for different publication types can be written. The examples presented in the table represent one style of Harvard referencing (see also Pears and Shields, 2005).

If you are writing for assessment purposes, then the referencing style you use may be a matter of personal preference or it may be dictated by the guidelines provided by your programme of study. It is common practice for programmes of study to recommend that descriptions of references are written out in full; that is, to include the title of the article. If you are writing for a particular journal, then the journal will insist that you use their journal style, which may or may not include the article title and may use full or abbreviated forms of journal names. Before using any particular style of referencing it is best to find out what the rules are, so as to avoid having to modify the referencing style later on. If you have to reformat your reference list to conform to the style of a particular journal, this is relatively easy to do if you are using reference management software.

You should know that journal abbreviations follow a standard system. If you are unfamiliar with the standard abbreviation used by a particular journal, then you can check the abbreviation through an electronic database such as Web of Science (Chapter 5). Some basic rules are

- One-word titles are not abbreviated but used in full (e.g. *Science, Biochemistry*).
- Journal is abbreviated to *J*.
- Titles ending in *-ology* are abbreviated to the *l* (e.g. *Toxicology* is abbreviated to *Toxicol*).

TABLE 1.2 How to write complete references for different types of publication in a reference list

Publication type	Information required and order of arrangement	Example
Book	Author(s) surname(s) and initials Year of publication (in parenthses) Title of the book (in italics) Edition (if not the first edition), abbreviated to 'edn'. Publisher's name Place of publication	Brown, T. A. (2004). *Gene cloning and DNA analysis,* 4th edn. Blackwell Publishing, Oxford.
Chapter in an edited book	Author(s) surname(s) and initials Year of publication (in parenthses) Title of the chapter Surname(s) and initials of editor(s) Title of book (in italics) Edition (if not the first edition), abbreviated to 'edn'. Publisher's name Place of publication	Mason, P.J., Enver, T., Wilkinson, D., Williams, J.G. (1993). Assay of gene transcription in vivo. In: Hames, B.D. and Higgins, S.J. ed. *Gene transcription: a practical approach.* Oxford University Press, Oxford.
E-book	Author(s) surname(s) and initials Year of publication (in parenthses) Title of the book (in italics) Edition (if not the first edition), abbreviated to 'edn'. Publisher's name Place of publication Type of medium (in square brackets) URL Date when last accessed by you (in parenthses)	Haines, J.L., Kort, B.R., Morton, C.C., Seidmen, C.E., Seidmen, J.G., Smith, D.R. (2007). *Current protocols in human genetics.* Wiley Interscience, UK. [e-book]. Available at: http://0-www.mrw.interscience.wiley.com.wam.leeds.ac.uk/emrw/9780471142904/home (last accessed 21 May 2007).
Journal article	Author(s) surname(s) and initials Year of publication (in parenthses) Title of article Name of journal (in italics) Volume number (in bold) Issue number (in parenthses) (if relevant) Page numbers	Dawo, M.I., Wilkinson, J.M., Sanders, F.E.T., Pilbeam, D.J. (2007). The yield and quality of fresh and ensiled plant material from intercropped maize (*Zea mays*) and beans (*Phaseolus vulgaris).* *J. Sci. Food Agric.*, **87,** 1391–1399.
Journal article that is available online but is not yet in an issue	Author(s) surname(s) and initials Title of article Name of journal (in italics) Published online Date of publication Digital Object Identifier (doi)	Schmutz, S.M. and Berryere, T.G. The genetics of cream coat color in dogs. *J. Hered.*, published online May 7, 2007. doi:10.1093/jhered/esm018.

TABLE 1.2 Cont'd

Publication type	Information required and order of arrangement	Example
Article in an e-journal	Author(s) surname(s) and initials Year of publication (in parenthses) Title of article Name of e-journal (in italics) Volume number (in bold) Issue number (in parenthses) (if relevant) Page numbers (if available) URL When the material was last accessed by you (in parenthses)	Garrick, D. (2008). Body surface temperature and length in relation to the thermal biology of lizards. *Bioscience Horizons,* **1,** 136–142. Available at: http://biohorizons.oxfordjournals.org/(last accessed 8 June 2008).
Conference papers	Author(s) surname(s) and initials Year of publication Title of article Name of conference Location of conference Dates of conference Publisher and place of publication Page/abstract numbers	Walker, O. and Conner, M.K. (2007). Molecular links between obesity and breast cancer. Experimental Biology 2007 Annual Meeting, Washington, DC, 28 Apr–2 May 2007. *Faseb J.*, **21**(6), A927–A927.
Theses and dissertations	Author's surname and initials Year of publication (in parenthses) Title of thesis/dissertation (in italics) Degree award in brackets, e.g. [PhD] Name of university awarding the degree Location of university	Divan, A. (2000). *p53, life, death and differentiation in retinoblastomas.* [PhD]. University of Sheffield, Sheffield.
Internet sites[a]	Author(s) surname(s) and initials (or corporate author) Year of publication (or when page was last updated) (in parenthses) Title of document (in italics) Medium of publication (in square brackets) URL When the material was last accessed by you (in parenthses)	WHO (2007). *Avian Influenza: situation in China* [online]. Available at: http://www.who.int/csr/don/2007_03_01a/en/index.html (last accessed 5 March 2007).
Audiovisual material	Author(s) surname(s) and initials (or corporate author) Year of publication (in parenthses) Title of publication Medium of publication (in square brackets) Name of publisher Place of publication	The Open University (2007). *In search of syphilis.* [CD]. The Open University, Milton Keynes.

(continued)

TABLE 1.2 Cont'd

Publication type	Information required and order of arrangement	Example
Reports (technical and official)	Author(s) surname(s) and initials (or corporate author) Year of publication (parenthses) Title of publication (in italics) Identifying letters and numbers of publication Publisher's name Place of publication	International Organization for Standardization (1972). *Documentation International code for the abbreviation of titles of periodicals.* ISO 4:1972. ISO, Geneva.
Patents	Name of patentee Title of patent (in italics) Country that granted the patent Patent specification number Date patent granted	University of Leeds. *Nucleic acids nematicides*. GB patent specification EP1780264. 2007.
Personal communication	Personal communications such as e-mail correspondence or telephone conversations should be incorporated in the main text at the point you refer to them as follows: (Zhang, personal communication). If you are citing unpublished data, the format in-text is (Horst, unpublished) or if the manuscript is in prepration (Graham *et al.,* in preparation). Personal communications, unpublished data, and manuscripts in preparation (but not yet accepted for publication) are not listed in the reference list.	

Note: the examples presented in the table represent one style of Harvard referencing only. Specific details may differ according to the instructions provided by your institution or journal: check the instructions before you start to write.

[a] There is no agreed system for referencing web-based sources. However, it is recognized that in addition to the URL, you must include details of when the web page was last updated and when you last accessed the material.

Referencing guidelines

1. You should always use high-quality sources of information which are factually correct, up to date, and reliable. Your major source of information should be peer-reviewed scientific articles published in journals. However, authoritative information such as patent details, government statistics, research legislation and manufacturer's protocols are also appropriate sources of information (Chapter 4).
2. You should cite only those references in your work that you have actually read. For example, if you have read a review article by *Jones* that discusses the work of *Brown*, then you should cite *Jones* as your source and not *Brown*. It is good practice, however, to read the primary article instead of relying on the interpretations of a secondary article when writing your work.
3. You should, as far as possible, cite primary published literature in your work. However, in cases where a comprehensive review article is available, then citation of the review article may be preferable to citing many separate references.

4. When selecting references, consider how relevant they are and include only those that are necessary to understand your work.
5. You must acknowledge appropriately all the material used in your work. This means citing all the sources both in the text (at the point you refer to the source) and in a list at the end of the writing.
6. You must use a single style of referencing consistently throughout your work. If you are writing an assignment for assessment purposes, then it is likely that you will be provided with guidelines on which is the preferred method to use by your tutors. If you are writing for publication in a journal, then you will need to check the journal's instructions for authors to see which style is acceptable to the journal you are aiming to publish with.
7. You must cite all your references completely and correctly, both within the text and in the reference list at the end. This means checking that all the references cited in the text are included in the reference list at the end. It also means checking the details of individual references against the original publication to make sure the reference details are correct as are all journal abbreviations, arrangement of each component of the reference, and the use of full stops, commas, spaces, italics, and bold type.

1.3 The writing process: planning, writing, and revising

Good writing is not only accurate, concise, clear, and referenced, but it should also be complete in content and organized so that it shows a logical progression of ideas. To achieve this, you should give yourself **sufficient time** to **plan**, **write**, and **revise** you work. These three stages are common to all forms of communication and are described in turn below. Each of these stages will be revisited in the context of the different types of communication in relevant chapters that follow.

1.3.1 Stage 1: planning your work

Planning a piece of work before you start to write makes the process of writing much easier. It does so by helping you to identify the key points to include in your work and in working out the order in which you can introduce these points. There are a number of techniques you can use to facilitate the planning stages of your work. These include **brainstorming**, the **mind map**, and the **linear plan** (Table 1.3).

What form of planning you use is a matter of personal preference. You may choose to use one of the techniques described here or other methods that you may already be familiar with. In any case, it is worth spending some time on the planning stage as it will ensure you do not omit key points and your writing is logically organized. This will make your work easier to follow and therefore more comprehensible to the reader or listener.

TABLE 1.3 Planning techniques: brainstorming, mind map, and the linear plan

Brainstorming

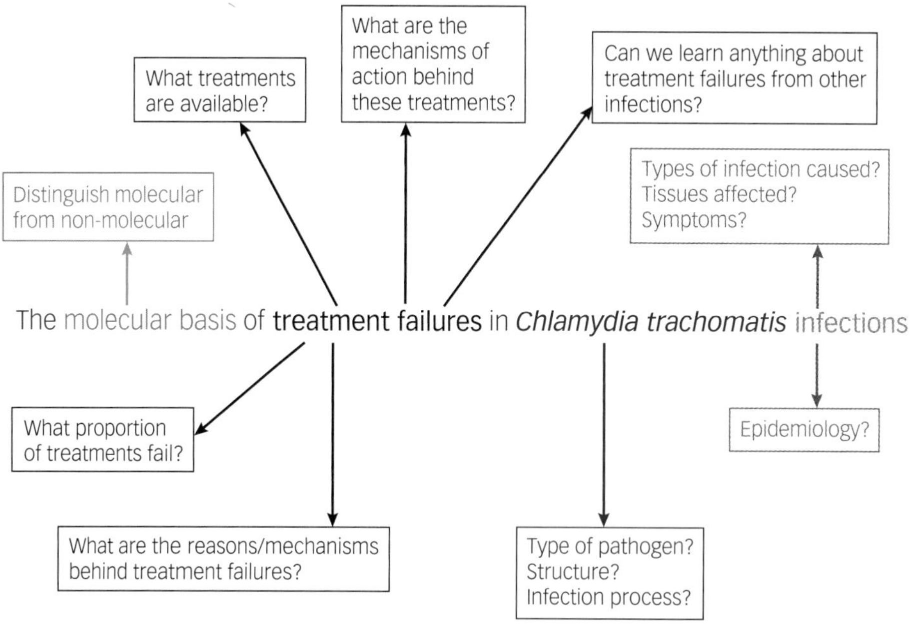

A simple technique which is often used at the early stages of planning to identify material that may be relevant for inclusion in your work. The aim is to get down on paper your main ideas. In a brainstorming map, the central topic is placed at the centre of the paper and then ideas as they come to you are written down in any order around the central topic. An example of a brainstorm is shown above and in Figure 7.2.

Mind map

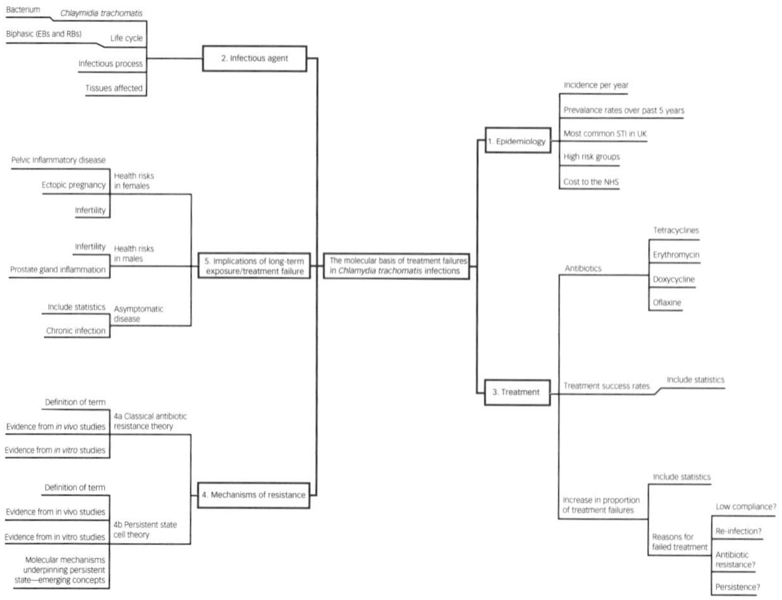

TABLE 1.3 Cont'd

A technique developed by Buzan (1991) which can be used to plan an overview of a topic area and identify relationships between the sub-topics. In a mind map, the main topic is written in a box in the centre of a page. Then, working in a clockwise direction, the sub-topics are written, again in boxes, branching from the centre. Further sub-points then branch off from the main points. Mind maps can be generated by hand, or there are a number of software packages available such as MindGenius that can be used to produce the maps electronically. An example of a mind map is shown above and in Figure 7.3.
Linear planning
A much more traditional form of planning in which the main points are listed sequentially under a series of sub-headings. An example of a linear plan is shown below and in Box 7.1.

Review topic: The molecular basis for antibiotic treatment failures in *Chlamydia trachomatis* infections

Introduction

- Establish importance of the topic (increasing prevalence, increased treatment failures infections and associated health risks)
- Delineate the review content: mechanisms underpinning treatment failures (antibiotic resistance and persistent state cell theory)

Main body

I. Epidemiology and pathogenesis of the disease
   - Causative organism
   - Most common STI in the UK
   - Increasing prevalence in the UK
   - High risk groups
   - Cost to the NHS to treat infections
   - Life cycle of *Chlamydia trachomatis* and survival in host cell.........

1.3.2 Stage 2: writing a first draft of your work

Once the planning is complete you can begin to write a first draft of your work. You do not need to write a piece of work from start to finish. You can start with any section that seems the easiest. You could also add sub-headings and then add details below each sub-heading. This can facilitate the writing process, and any sub-headings that you do not require can be removed later during the redrafting stage of your writing. The aim of the first draft is to produce a tentative outline of your work which can be refined and fleshed out with details as you progress through your writing. When writing the first draft, you will need to take into account any guidelines you have been supplied with on the structure of the report or presentation and the word limit. For example, research papers follow a standard IMRAD structure in which the **I**ntroduction, **M**ethods, **R**esults and **D**iscussion are presented sequentially. A literature review, on the other hand, is structured into three conventional parts: the introduction, the main body, and the conclusion, with or without sub-headings. You should therefore ensure you are familiar with any guidelines provided and conform to them exactly.

When writing you may experience 'writer's block', during which your writing stalls. If this happens, do not worry: it is common to most writers. When this happens, try the following techniques:

- Take a break from your writing. You can then return after a break refreshed and ready to continue.
- Work on a different section of your writing. Working on different parts simultaneously can sometimes help clarify other parts and facilitate writing generally.
- Talk to someone about your work. This can also be useful in assisting you in developing your ideas.

1.3.3 Stage 3: revising your first draft

You will need to refine your first draft a number of times before you reach a final copy that is accurate and complete and ready for submission or presentation. When redrafting, do not start revising immediately after you have completed your first draft. Instead, leave it for at least a day and then come back to it and start revising. When revising you need to check:

- The accuracy and completeness of the content of your work.
- The organization and coherence of your writing including spelling, grammar, and punctuation.

Checklists for reviewing your work are included in the relevant chapters that follow. As you refine each successive draft, it will help if you focus on different elements of your writing. For example, the first time you redraft you may focus on the content and rearrange, delete, or add further text. Subsequently you may focus on checking the accuracy of the references. On further readings you may decide to focus on the spelling, grammar, and punctuation. With each successive redraft your work will improve and you will end with a professional final version that is ready for submission or presentation.

1.4 Interpersonal communication: working with your supervisor and as part of a research group

Interacting with and building productive working relationships with your supervisor, your research group members, and other scientists require good communication skills. This section provides some guidance on how to manage your relationship with your supervisor and with members of your immediate research group. Developing and maintaining contact with other scientists working in a similar or related field is discussed in Chapter 15 on networking.

1.4.1 Working with your supervisor

To build a strong working relationship with your supervisor you must first understand what you can realistically expect from your supervisor and what your supervisor expects from you. These expectations will be outlined in your university's code of practice for students. Typically, the role of the supervisor is to **support** and **guide** you through your research and in all aspects of the training surrounding that research. This typically includes helping you define your research question and work plan; approving and making sure you follow a timetable of work; and discussing results and commenting on drafts of your thesis. It also includes assisting you to identify appropriate training such as health and safety and transferable skills. Your supervisor will expect you to work hard, meet deadlines, be independent and show enthusiasm for your work. Some guidelines to help you get the most out of your relationship with your supervisor are:

- Find out what you can expect from your supervisor and what your supervisor expects from you. This means reading your university's code of practice for students and any departmental and research group guidelines. It also means speaking to your supervisor(s) and clarifying what you can expect (and what they expect). By the end of this process, you should have a clearer understanding of your roles and of the practicalities, such as how frequently you can expect to meet with your supervisor on a one-to-one basis, the procedure for contacting your supervisor, and the frequency and format of research group meetings.
- If you have more than one supervisor, then find out what the individual responsibilities of each supervisor are. The following are the types of information you should know. Which supervisor is your first point of contact? Will you have regular meetings with both (all) supervisors or one supervisor? How will you keep the second supervisor(s) informed of your progress?
- Be professional in your relationship with your supervisor(s). This includes attending meetings on time, informing your supervisor if you can't attend a meeting, taking responsibility for scheduling meetings, and completing work according to agreed deadlines.
- Prepare carefully for meetings with your supervisor. Always have a clear agenda of what you will discuss at the meeting and what you expect to achieve by the end. If you are meeting to discuss the progress of your work, then take your data with you (both raw and analysed) and be prepared to describe and explain your findings. Share all the data that you have accumulated since the last meeting and highlight any inconsistencies in your data, any technical problems you've experienced, how you've overcome them, and any other information that will accurately convey the progress of your work. Always keep a record of the meeting, and in particular any objectives you have agreed for further work. This will help to monitor your progress, and the record can be used as evidence that a particular discussion took place if any problems arise at a later date.
- Expect your supervisor to give you critical (but constructive) feedback on your work. Do not become defensive when this happens. The aim is to improve your

understanding and quality of your work. Therefore, consider the feedback you receive carefully and use it to enhance your work.

- You will communicate with your supervisor by e-mail often and for many purposes. These could include arranging meetings, describing how your work is progressing, asking questions, and sending drafts of your work. To communicate effectively by e-mail, follow the guidelines provided in Chapter 2 (section 2.2).
- Your supervisor will expect to see that you are developing stronger analytical and practical skills as your work progresses. Therefore, you should expect your relationship with your supervisor to mature and evolve over the course of your research. This will be reflected in you becoming more independent and less reliant on your supervisor for advice. It could also mean that you meet less frequently on a one-to-one basis to discuss your work.

Most people develop a productive relationship with their supervisor which continues after the programme of study has come to an end. However, if you have concerns about your relationship with your supervisor, then discuss your concerns with your supervisor or seek advice from the support structure that your department has in place for supporting students. This could involve, for example, talking to an advisor (or other identified person) allocated to mentor you during your programme of study.

1.4.2 Working as part of a research group

Research groups are made up of people working on different projects within a particular topic area or on different aspects of the same project. These groups typically consist of PhD and Master's students, postdoctoral research fellows, and technical staff, and are headed by a team leader (also termed the principal investigator or supervisor).

When you first enter a research team, it is important that you learn the ground rules. The following are the types of information you should know. What are the routine jobs and how are they divided amongst the group? Which resources are shared by the group members and which are for personal use? What is the procedure for ordering consumables and booking equipment? Which areas of the lab are for common use and therefore should be cleaned up after use? By contributing to routine laboratory or group duties and by working in a way that does not damage or contaminate someone else's work, you will establish a reputation for being a responsible member of the group. This will also minimize the possibility of tensions arising between you and other group members.

You should also understand the roles of each group member. For example, if there is a senior postdoctoral research fellow in the group, s/he may be responsible for overseeing the daily activities of the more junior researchers or there may be a designated individual for overseeing the booking system of particular equipment. Understanding the roles and responsibilities of each group member will help you respect the role of others and assist you in establishing your role within the group.

In addition, find out the particular area each member in the team is working on (attending research group meetings (see section 1.4.3) will help you learn about the projects of other group members). This will help you to see how your work fits in with the wider group and you will know who to approach for advice and information when you require it.

You should also understand the culture of the group. In most research groups, each member works on an independent project but cooperates with others in sharing what s/he knows with other members of the group (and with different research groups within the same department or institution). Sharing of information can include recommending literature to read, sharing protocol information, technical expertise, and helping in the interpretation of data. This cooperation between group members is important as it increases research output. Team members can also collaborate with each other on projects (and with other researchers in different groups or institutions) to work on aspects of the same project. Team work can raise complex issues around ownership of data, authorship, and acknowledgement of contributions made by others. These issues are discussed in Chapter 3 as part of the ethics of communication.

On the whole, working as part of a research team is an enjoyable experience and provides you with a support network of like-minded colleagues with whom to share information and experiences. Of course, there can be instances when friction arises between group members. Friction could arise, for example, if a highly competitive member of the group is less willing to share information, or a member fails to contribute equally in routine laboratory duties, or the contributions of a member to a particular project are not acknowledged. Conflict between group members will undermine the effectiveness of the group and it is therefore important that any problems are resolved quickly. It may be possible to resolve some types of conflict by the group members working through the issues themselves. Other types of conflict may require intervention from the team leader. Do not be afraid to ask for help if you are experiencing a situation you have difficulty in handling.

1.4.3 Research group meetings

Most research groups meet periodically, usually weekly, to report their latest research findings to the rest of the group. There are many advantages to the group meeting. One is that it will keep the team leader informed of your work. The other is that in order to present your findings to the rest of the group you will need to analyse and interpret your data, which will help you understand the data and plan subsequent experiments. In addition, your work will benefit from the input of others in the team who will ask you questions about your experimental strategy and the results generated. This will assist with troubleshooting experiments if you are experiencing any difficulties with protocols. A further advantage of the meetings is that you will begin to develop your presentation skills and your ability to receive and give constructive criticism. These skills are a necessary part of presenting your work at scientific conferences (Chapters 13 and 14) and when defending your work at a viva (Chapter 12).

Discussions at research group meetings can be lively and informative. To benefit fully, be prepared to present your latest data and take an active interest in the work of other members of your group. Take your raw and analysed data with you to the meeting and be prepared to describe and explain your findings. Before the meeting, anticipate the questions you may be asked and think about what your responses could be.

At the meeting, you should expect your group members and supervisor(s) to ask you critical question about your work. The aim will be to establish whether you have conducted the work rigorously and generated data of a high quality. This critical approach to questioning is an essential part of evaluating scientific literature (described in Chapter 6) and is intended to improve the understanding and quality of your work. Therefore, do not become defensive when your work is subjected to critical scrutiny by your peers and supervisor. Instead, listen carefully to the questions asked and provide well-thought-out and reasoned responses. If you do not know the answer to a particular question, then it is entirely acceptable to admit you don't know. However, your inability to answer questions should not be a consequence of poor preparation or a careless attitude towards your work. Overall, you should demonstrate to your supervisor (and to the team as a whole) that you have made an attempt to understand your work and considered what the next stages of your experiments will be.

When giving feedback to your peers on their work, be constructive. This means identifying both the negative and positive aspects of the work and, where the work is weak, outlining the evidence that supports your opinion. The feedback is more likely to be received positively if it is given in this fair and objective manner. Remember, the aim is to help others improve on their work and this is only possible if the feedback is honest and the environment in which it is delivered, supportive.

References

Buzan, T. (1991). *The mind map book*. Penguin, New York.

CSE (2006). *Scientific style and format: the CSE manual for authors, editors, and publishers*, 7th edn. Council of Science Editors in cooperation with The Rockefeller University Press, Reston, VA.

Pears, R. and Shields, G. (2005). *Cite them right: the essential guide to referencing and plagiarism*. Pear Tree Books, Newcastle upon Tyne.

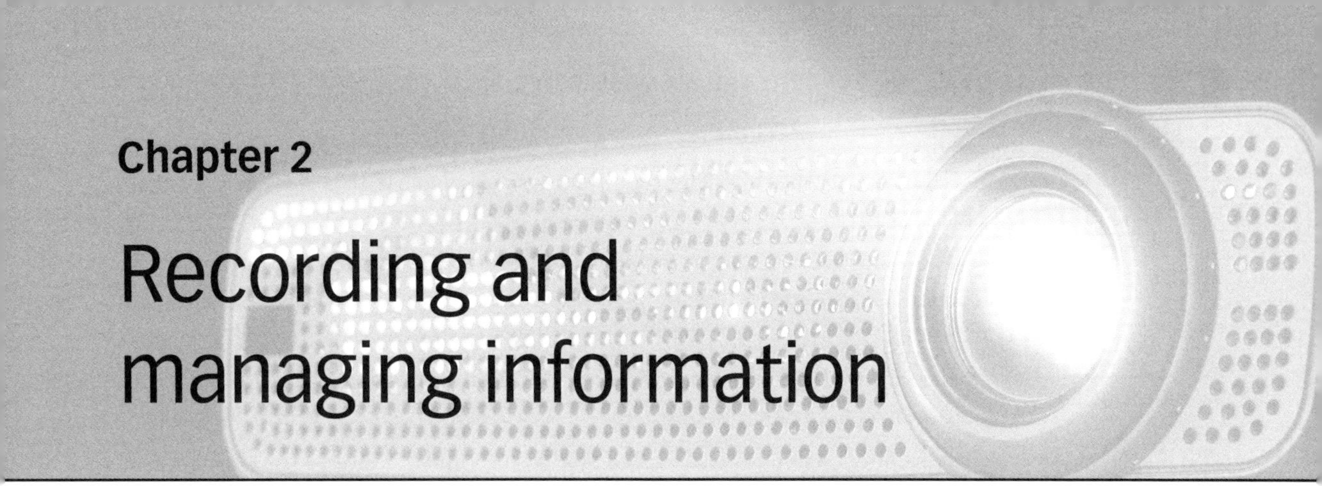

Chapter 2

Recording and managing information

➲ Introduction

This chapter provides guidance on recording and managing the large amounts of information you will build up during the course of your programme.

Specifically, this chapter will:

- provide guidance on writing good experimental notes of your research work
- provide guidance on how to manage your e-mail account efficiently
- provide guidance on how to organize and manage your electronic and hard copy information.

2.1 Maintaining experimental records of research data

During the course of your research, you will conduct a large number of experiments and generate a substantial amount of experimental data. You will need to maintain accurate and detailed records of these experimental procedures and the data generated. This includes raw data such as photographs, questionnaires, and field work observation log sheets, and the analysed and processed data such as statistical spreadsheets and graphs.

Your experimental notes serve a number of valuable purposes:

- Your notes will help you to write up your work accurately for any dissertation or thesis that you write for assessment purposes or any papers that you prepare for publication.
- Your notes will help you establish that your research has been conducted honestly and your results and their interpretations are authentic if the integrity of your conduct or your results are challenged at any stage.
- Your notes are a legal document which is used as evidence to support a patent application, if you make a patentable discovery.

To ensure that experimental notes are fit for the purposes described, all experimental notes should meet the following requirements (see also Medical Research Council, 2005):

- Your experimental notes must include a **complete** record of each experiment that you conduct, including the aim or hypothesis, the materials, procedures, results, and conclusions.
- Your experimental notes must be sufficiently **detailed and comprehensible** so that another competent scientist working in the same field can reconstruct your work exactly.
- Your experimental notes must be **accurate**. You should therefore write your notes as you are conducting your experiment, or write them up as soon as possible after the data are collected.
- Each entry should be **dated** so that there is a clear trail of when each experiment was conducted and when the data were collected.
- Experimental notebooks should be **checked and signed** by your supervisor(s) regularly to authenticate your notes as an accurate and complete record of your research.
- Experimental notebooks and the associated primary and analysed data should be **kept securely** so that they are not damaged, lost, stolen, or tampered with during the course of the research.
- Completed experimental notebooks and the associated data should be **retained** for a period of time in a secure location after the work is complete. The minimum time period for retaining research data will depend on institutional and funding agency policies. For example, the Biotechnology and Biological Sciences Research Council expects that data generated through its funding should be kept for a period of 10 years after the work is complete (BBSRC, 2007).

Your experimental notebooks are the property of the institution which received the funding to support your project (Chapter 3, section 3.2). This means that you will not be able to take your experimental notebooks with you when you leave the institution unless you are granted permission to do so by your supervisor.

2.1.1 Maintaining experimental notes

Records of experimental work are typically maintained as a hard copy in bound notebooks. These records include the aims of the experiment, materials, experimental procedures, results, and conclusions. The results include primary raw data as well as analysed and processed data. Some of these data may be computer-generated and can be printed out as a hard copy (such as photographs and chart recordings), but some may be available in electronic format only (such as audio and video recordings). Your experimental notebooks are therefore likely to consist of a combination of hard copy notebooks and electronic stores of data. You will need to organize this information carefully so that your experimental records are comprehensible to another scientist looking through your work.

Practical guidelines

- Use a hardback notebook for writing your experimental notes. Number each page consecutively, write your name at the top of each page, and date each page or entry. When dating use a format such as 21 Oct 2007, rather than 21/10/07. This is to avoid confusion with different date formats that are used in different countries.
- Use a ballpoint pen or other permanent pen for all entries. If you make a mistake, do not erase or scratch it out but instead put a single line through the mistake so that it still remains readable. If you make any changes, then initial and date the change, and explain briefly why the alteration was made. If you leave any blank pages or part-blank pages in your notebook then put a diagonal line through from corner to corner to show the page is blank.
- Although it is normal to make entries in permanent ink, there may be occasions when the use of a pencil to record notes is acceptable or even preferred. An example is when you are conducting field work and the surface you are writing on is wet; as a pencil writes better on a wet surface than a permanent pen.
- When writing up each experiment include the following sections:
    - The **title of the study** and a **brief description** of the experimental aims or hypothesis.
    - **Materials:** this should include details about subjects (e.g. names of cell lines and their passage numbers), special reagents (e.g. antibodies, cell culture medium), and special equipment (e.g. microscope including type, serial number, supplier) used. It is important to note down reagent suppliers and batch numbers where available. In the case of field studies, include information such as the exact site location, description of the physical and ecological characteristics of the site, and environmental conditions under which the experiment was conducted.
    - **Procedure:** this should include a step-by-step description of your experimental protocol. If you are using a protocol that is already written down, for example in a methods file in the laboratory or in your student handbook, do not write it out again but instead reference the source. However, if you make any modifications to the protocols contained in a methods file or student handbook, then you should document the exact changes in your notebook.
    - **Results:** this section should include your raw data and your analysed data. Hard copies of raw data (such as photographs and spectra) should be taped directly into your notebook. When attaching data in this way, make sure you date and annotate them sufficiently so that you can identify what they are and where and how they were obtained. If it is not possible to insert the data directly into a notebook because it is too bulky or it is in electronic format only (e.g. video recording), then store the material in a ring binder (for hard copy material) or on a server and storage device (for electronic material) and index clearly where the data are stored.

- **Discussion:** this is a brief summary of your conclusions. If the experiment did not work, include a statement on what went wrong and what you will do next time to troubleshoot.

- Back up all electronic copies of your data regularly (see section 2.3) and make routine photocopies of your hard copy notes. Keep the duplicate copies in separate locations.
- Keep copies of both your processed and unprocessed data. For example, if any images (such as gels, immunoblots, immunohistochemistry) have been electronically

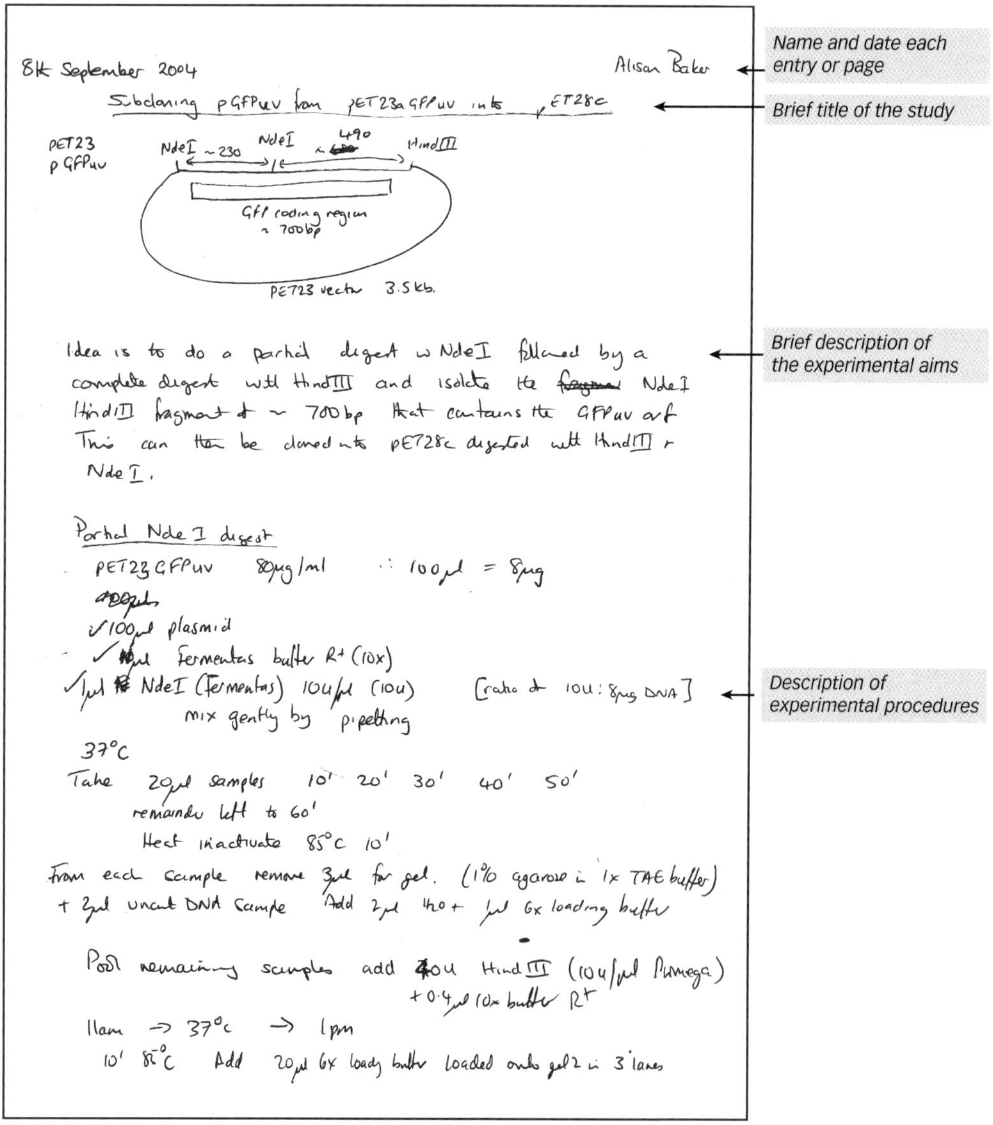

FIGURE 2.1 Examples of experimental note pages from a handwritten experimental notebook. The two pages represent a complete experiment: including the title, aim, methods, results, and discussion of the study. Both pages are annotated to highlight the types of information to include when writing up each experiment.

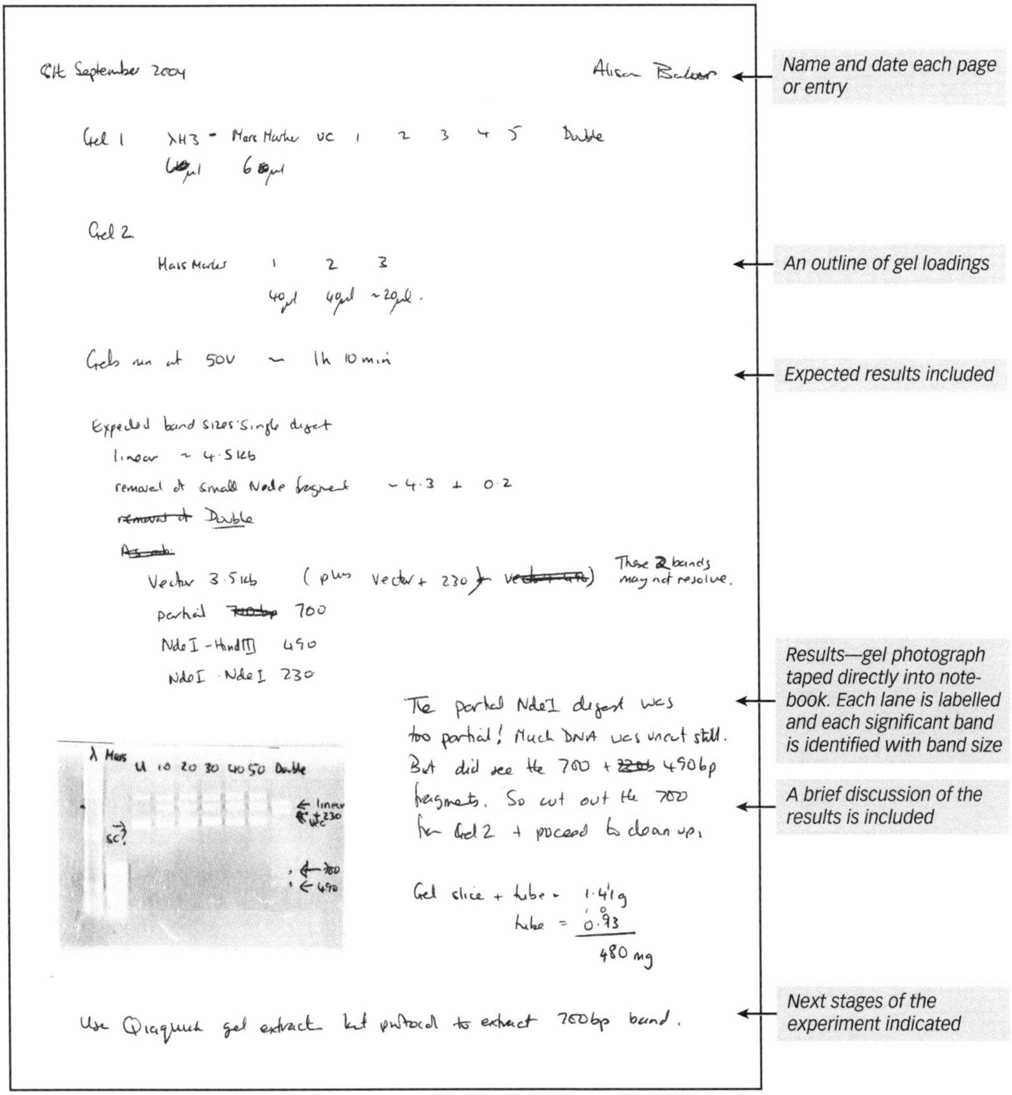

FIGURE 2.1 Cont'd

processed, then keep copies of both your processed and unprocessed data and make a note of which software and version you used to process the data and the exact details of how the image was processed (see Chapter 3, section 3.3.3).

An annotated example of laboratory notebook pages is included in Figure 2.1. You can also view examples of experimental notes and personal papers by some of the world's leading scientists (such as James Watson) at Genentech Center for the History of Molecular Biology and Biotechnology (http://library.cshl.edu/GCHMBB/index.html). This is an archive established by Cold Spring Harbor Laboratory and funded through the biotechnology company Genentech.

2.1.2 Electronic notebooks

Although it is typical to maintain experimental notes as a combination of hard copy notebooks with stores of cross-indexed electronic data, it is also possible to maintain your notes in a single location using an electronic medium. When you use this method, your notebook must meet the same requirements as the hard copy notebook: that is, there must be an accurate, detailed and reliable trail of your work which is authenticated by your supervisor. There are software packages, termed electronic laboratory notebooks or ELNs, which are commercially available and specifically designed to meet these requirements. Each entry is date and time stamped, and any changes made and who made them is recorded. You can return to earlier versions of the experimental notes and the notes can be authenticated by electronic signatures (Butler, 2005).

Maintaining notes electronically has some advantages over hard copy, particularly if your work involves large amounts of computer-generated data. Amongst the advantages are ease of recording and managing data sets, ease of making back ups of your work, search facilities, and the ability to share your notes with other members of your research group by granting authorized access. However, there are also significant limitations such as less portability of notes, malicious hacking, breakdown and corruption of computer systems, and the expense of purchasing a commercially available ELN. In addition, software formats change rapidly, which means that data may have to be converted to new formats to ensure accessibility for the period of time the data are expected to be retained. For comprehensive reviews of the advantages and limitations of ELNs, see Drake (2007) and Du and Kofman (2007). Although the use of electronic notebooks is expanding in pharmaceutical and biotechnology industries, their use in an academic setting is still limited (Nature, 2007). However, it is worth exploring the use of an ELN, particularly if your work generates large amounts of automated data.

2.2 Managing your e-mail: some simple guidelines

E-mail is an extremely useful tool for sending and receiving messages between individuals or between groups of people easily and quickly. You are already almost certainly making frequent use of e-mail for both work and personal use. This section lists some simple guidelines for using e-mail effectively and efficiently in your work.

2.2.1 Using e-mail

- Use your e-mail account securely, by logging in with a secure password which you change frequently (this applies to your computer as a whole).
- When composing an e-mail:
    - Add a **meaningful title** in the subject field to inform the recipient of the content of the e-mail. This will help the recipient identify the content of the e-mail, and store and search archived e-mails easily.

- Start the e-mail with a **greeting**. The type of greeting will depend on how well you know the recipient. For example, if you are writing to friends or colleagues you know well, then your greeting can be informal 'Hi . . . .' If you are writing to people that you know less well, such as your tutors, then your greeting should be formal, for example, 'Dear Dr. . . .'
- Match the **length** of the message to the purpose of the mail. For example, if you are asking a question, keep it short and focused. Rather than sending a long e-mail covering multiple topics, you should consider whether individual e-mails for each topic area would be more suitable than a single long message covering multiple topics.
- Use **spelling, grammar, and punctuation** correctly. Do not use text language (e.g. thnx, u, or plz) unless you are writing to your friends. You should also avoid using capital letters as this is considered shouting in electronic forms of communication.
- Finish your e-mail with a **farewell term** such as 'best wishes', 'thank you', or 'look forward to hearing from you' and sign off with your name and surname. You could also include your contact details (address and telephone) and any other information that may be useful to the recipient to help identify you (such as a student identification number). If you create a suitable e-mail signature then your contact details will be added automatically to every e-mail that you send.

- If you are attaching files for transmission by e-mail, then consider the file size. Some files, particularly those which contain images, are very large, and this will slow down the speed of transfer of the message, or on some systems may block it completely. The best way of sending large files electronically is to use file transfer protocol (FTP). To transfer files by FTP, deposit the files on an FTP server and then send the URL link to the recipient with details of how to access the material. If you send the file as an e-mail attachment, then reduce its size by zipping multiple files together using software such as WinZip. If you are sending an image file you can reduce the size by compressing it (see Chapter 11, section 11.3).
- If you receive a file attached to an e-mail, never open it directly by double-clicking on it. Instead, save the attachment to a folder on your hard drive and then scan the file using your anti-virus software. If no virus alerts are seen, then it is safe to open the file.
- When replying to an e-mail, include the relevant sections from the original message so that the recipient can place your e-mail in context. This is particularly important if you are responding after a period of time.

2.2.2 E-mail etiquette

- Do not re-send the same e-mail multiple times to the same recipient. Most people receive many e-mails daily and the recipient is likely to get annoyed that you are clogging up their inbox unnecessarily.
- Do not expect an immediate response. Remember, if you are communicating with people in different time zones they may not be at work, or they may be working on other more important projects.

- Do not mark an e-mail with the 'high importance' tag unless the message is important and urgent. Again, remember, it may be a priority for you but not for the recipient and hence you may not receive an immediate response.
- Use the tracking of sent e-mail facility (e.g. message has been read by the recipient) sparingly. For example, request such a read receipt if you are unsure whether the recipient is reading their mail.
- When you send a message to multiple people, consider who really needs to be copied in. Similarly, if you receive a message that has been sent to multiple recipients, consider whether your reply should be sent to the sender only or to all of the people to whom the message has been sent. People's mailboxes can get cluttered by e-mails they have unnecessarily been copied in on.
- When sending an e-mail to multiple people, consider whether you should blind carbon copy (BCC) the e-mail addresses of the individual recipients. This function does not disclose e-mail addresses to other recipients receiving the same message and is therefore very useful when you are sending a single e-mail to multiple recipients but need to maintain the privacy of e-mail addresses.

2.2.3 Managing e-mail efficiently

The advantage of e-mail is that it increases the speed and ease of communication. However, the downside of this is that it can also increase the volume of e-mails you receive in return. Going through these e-mails can be time-consuming and it is therefore useful to have a system for responding to e-mails, and a system for locating and retrieving any important e-mails quickly. Here are a few tips to help you manage your e-mail account efficiently:

- Set up **filters** to deal with spam e-mail. Spam is unsolicited e-mail which is received from spammers (people and organizations) advertising products or sending opinions you do not want to hear about. To minimize the amount of spam e-mails you receive, your institution may already have set up an automated filtering system which filters spam (and other junk e-mails). If not, you can set up rules on your e-mail software to filter unwanted messages directly into a separate folder in your mailbox, as they arrive. This will not stop spam from arriving, but will prevent it clogging up your inbox.
- Read your e-mails **frequently** to avoid a build up of e-mails and to avoid missing any important or urgent information you may be sent.
- Deal with your e-mails on a **priority** basis:
  - Delete any spam e-mails.
  - Archive the e-mails you want to keep but which do not require a response (see archive bullet point below).
  - Deal with any urgent e-mails immediately.
  - Schedule e-mails which do not require an immediate action for a later date. You can do this by marking the e-mail by a follow-up flag and assigning a date by which the e-mail should be followed up.
  - If there is anyone else who should be responding to the mail you have received, direct it to the appropriate person.

- You will find that if you keep all your e-mails in your inbox, trying to locate an important message at a later date will become difficult. You should therefore **archive** all your important (sent and received) e-mails into folders. For example, you could create two main archive folders, one named 'inbox' and the second named 'sent items'. You can then create sub-folders within the main folders and archive your messages by either sender or topic. This will help you find an e-mail quickly if you need to refer back to a particular e-mail for information or as evidence that a particular discussion took place.

2.3 Organizing electronic and hard copy information

It is essential to set up a system to organize and manage the large amount of electronic and hard copy information you will build up during the course of your work. This information includes electronic copies of your word-processed documents, data files, graphics files, references, and presentation material. It also includes any useful websites you may access regularly. Hard copy information will include print-outs of your word, data, graphics, references, and presentation documents.

If you do not set up a system for managing your information, you will become overloaded and waste time trying to find things or loose important information and then waste time rewriting the lost material. To avoid this, follow these guidelines to organize your information:

- **Set up a system for storing and organizing your electronic files.** One way of organizing your files is to create a separate folder for each type of document (e.g. Word documents, Power-Point presentations, data files, and image files) and then create sub-folders within the main folders. You can then file your documents by document type and topic. Another way is to create folders by topic and then sub-folders by document type. Give both your folders and your files meaningful names and try to store each file in a single folder only. This will help you locate a particular file easily. If you are not able to locate a file easily, you can use the search command from the Windows Start menu to search for the file on your computer.
- **Set up a system for storing and organizing your electronic references.** Use reference management software such as EndNote, Reference Manager, or Procite to store your references (see Chapter 1, section 1.2). These management systems have a very useful function of allowing you to attach an electronic copy of the reference to the citation details stored in the software. You may also want to consider storing your references online, on sites such as Connotea (http://www.connotea.org) or CiteUlike (http://www.citeulike.org/). You can add your references to these sites and then assign tags to the references to organize them by topic. You can either keep your references private or share them with other people registered on the same site. The advantage of online reference storage systems is that you can access your reference list from any computer which has an Internet connection. You can also search the references shared by others on the site by subject tag to identify any additional references that could be of use to you.

- **Set up a system for saving and organizing useful websites.** You can save any useful websites that you come across by bookmarking them so that you can return to the site again easily. You can bookmark web pages in your web browser, organized into folders by topic, or you can bookmark them into an online reference storage facility such as Connotea or CiteUlike described above.
- **Make backups of your work regularly.** You must make back-up copies of your work regularly. You can use a CD (compact disk) to store your work or a USB flash memory (memory stick). You should also keep copies of your work in different locations: for example, one on your home computer, another on your work computer, and then further copies on a storage device. Another very useful way of storing your files is to export them onto a secure online server such as Google Docs (docs.google.com/) (see Chapter 15, section 15.4.5). The advantage of using an online server is that you can access (and edit) your files from any computer with an Internet connection.
- **Before you print out hard copies of information, consider whether it is strictly necessary to do so.** The more information you print out, the harder it is to organize and to keep track of the different versions of the printouts.
- **Set up a filing system for organizing hard copies of your files.** One way of organizing hard copies of your word, presentation, data, or graphics documents is to file them—by document type or subject matter—in A4 ring binders, lever arch files, or folders (it is worth investing in a filing cabinet to organize and store your information in). Label each file or folder in an informative way so that you can retrieve the document you are looking for easily. Make sure all your printouts are annotated with the date and the electronic file name so that you can keep track of the different versions of a particular file and locate the corresponding electronic file easily.
- **Set up a filing system for organizing hard copies of your references.** Similarly, you can organize your references either by topic, author, article type, or year of publication in files or folders.

References

BBSRC (2007). *Statement on safeguarding good scientific practice.* [online]. Available at: http://www.bbsrc.ac.uk/funding/overview/good_practice.pdf (last accessed 10 June 2008).

Butler, D. (2005). Electronic notebooks: a new leaf. *Nature*, **436**, 20–21.

Drake, D. (2007). ELN implementation challenges. *Drug Discov. Today*, **12**, 647–649.

Du, P. and Kofman, J.A. (2007). Electronic laboratory notebooks in pharmaceutical R&D: on the road to maturity. *JALA*, **12**(3), 157–165.

Medical Research Council (2005). *Good research practice: recording the data.* [online]. Available at: http://mrc.ac.uk (last accessed 29 June 2008).

Nature (2007). Editorial: share your lab notes. *Nature*, **447**(7140), 1–2.

Chapter 3

Ethics in communication

➲ Introduction

As a research scientist you will design and conduct experiments, analyse the data generated, and communicate your findings to others through a variety of media. These media include informal discussions with your research group members and more formal presentations of data through oral and poster presentations, scientific papers, and thesis or dissertation preparation. As you progress through these activities, you will need to consider ethical issues relating to research and the communication of research. This includes topics such as data ownership, data sharing, and authorship and acknowledgement issues. This chapter introduces some of the ethical questions relating to communication which you will encounter as a research student, and provides a framework for understanding what is acceptable and what is unacceptable scientific conduct.

Specifically, this chapter will:

- outline key ethical issues relating to data ownership and data sharing
- define plagiarism and identify strategies for avoiding plagiarism
- outline the key ethical issues relating to publication of primary research data including duplicate publication, conflicts of interest, authorship, and acknowledgements
- provide a basic overview of copyright and how to use copyright material in your work appropriately.

3.1 Ethical communication: an overview

As a scientist you are required to be honest, objective, accurate, and complete in the way you design and perform your experiments and in the way you interpret and communicate your findings. If you work in a manner contrary to this, then you could be guilty of scientific misconduct. Scientific misconduct includes (Benos *et al.*, 2005):

- **Fabrication,** which is inventing fictitious data.

- **Falsification,** which includes altering data or experimental procedures, or withholding data in order to present a particular conclusion or to avoid explaining a result which could complicate your interpretations.
- **Plagiarism,** which is using the work of others and passing it off as your own without crediting the original author, or crediting the original author incompletely.

Misconduct can arise for a number of reasons. One reason is that you do not clearly understand what types of behaviour are classified as scientific misconduct. Another reason may be that you are under pressure to complete a piece of work or publish a paper, which leads you to rush the work and therefore make careless mistakes. In both these cases the misconduct is likely to be unintentional. However, the pressure to complete a piece of work or publish a paper may lead you deliberately to fabricate, falsify, or plagiarize a piece of work. The consequences of misconduct can be very severe: it may result in you being expelled from your programme of study, and it can have legal consequences. It is therefore essential that you work in a way that is ethically acceptable with full knowledge of your responsibilities as a research scientist—these responsibilities as they relate to communication are discussed in this chapter.

An added complexity is that research is almost always conducted in teams which include multiple people working on different aspects of the same project. These teams may be composed of people from the same research group or people from different research groups, different disciplines, different institutions, and even different countries. This multidisciplinary and collaborative approach to research clearly has benefits in terms of sharing information, expertise, and resources, which in turn accelerates scientific discoveries. However, it also raises questions around who owns the data and who should be listed as a co-author on publications (Ledford, 2008)—hence ethical issues around data ownership, data sharing, and authorship are also discussed in this chapter.

There are a number of resources you may want to consult on communication ethics. The Committee on Publication Ethics (COPE) (http://www.publicationethics.org.uk/) and the On-line Ethics Centre (http://www.onlineethics.diamax.com/) include informative case studies. In addition, there are useful publications on ethical conduct of research and its dissemination that are available through the Research Integrity Office (UK) (http://www.ukrio.org) and Office of Research Integrity (USA) (http://ori.dhhs.gov/) websites.

3.2 Data ownership

Research data that you generate as part of your research are generally the property of the institution which received the funding to support your project. Data include experimental notebooks as well as the resources generated, such as cells; sequences; computer codes and programmes; biochemical and chemical molecules; spectroscopic data; and plants and animals. Cases where ownership issues may be less clear is if your work is funded by an industrial sponsor or if you are part of a collaborative project with other institutions. In such cases you will need to check the ownership agreements made between the institution and the sponsor or collaborator by consulting your supervisor.

Although you do not own the data, you have the right to be acknowledged as the person who conducted the work and generated the data on any publications (abstracts, oral presentations, research papers) that originate from the work. Depending on the type and level of contribution to a research project, this may be in the form of co-authorship of the manuscript or as an acknowledgement in the acknowledgements section of the manuscript. Both these issues are discussed in section 3.5. In the event of your work being commercialized, then you may well get some form of financial compensation or revenue back as a result of the commercialization.

If you leave a particular institution, for example at the end of your research project, then you will need to ask permission before you can take a photocopy of your experimental notes with you. Likewise, if you plan to continue with a similar line of work and require access to some of your material such as a particular clone, then again you will require permission before you can to remove it from the institution. Remember, obtaining agreement to use the material does not mean that you own the material, but that you have been granted permission to use it for further research. The transfer of tangible research materials from one institution to another is made through a simple legal agreement called the **material transfer agreement** (MTA) which sets out the conditions under which the recipient can use the material.

3.3 Data sharing

The process of communicating your research work usually follows a sequence:

- The data you generate are initially presented and discussed at your research group meetings with your team members and supervisor(s) (see Chapter 1, section 1.4). If you are collaborating on a project with external organizations such as industrial partners or other institutions, then you will also share your data with your collaborators.
- As you build up your data, the next stage of dissemination is usually to present your work in poster or oral format at a scientific conference. Presenting at a conference is an important part of the communication process as it opens up your initial data for comment and discussion with the wider scientific community and informs them of your work.
- As your research progresses, you will build sufficient data to write a full paper and hence the next stage of dissemination is to publish a research paper. There are a number of benefits to publishing your work either as an abstract at a conference or as a full paper. One is that you will take credit for the finding and another is that you will begin to build up your publication record and hence your reputation as a scientist. A strong publication record will help you obtain research funding and secure jobs and promotions.

Before sharing your research data at conferences or as a research paper, there are two important questions you should consider.

1. When are the data ready for sharing?
2. How much data should you share?

Some considerations are:

- Some data are so preliminary that they are not at the stage where they can be described or evaluated and therefore are not ready for sharing with the public. This could be, for example, if you have not yet repeated and validated your findings.
- If the data you are generating could be patentable, then you will not be able to present or discuss them at conferences or publish them until after the work has been patented. Public disclosure prior to filing a patent will mean you can no longer patent your work (in the UK: but law may vary in different parts of the world). You should take advice from your institution's intellectual property office or legal department as soon as possible if the data you are generating could be patentable.
- If the information is private, such as that relating to human subjects: you should take care not to disclose the private details of any of your subjects.
- You may be restricted from sharing your data with others if the contractual conditions of your funding or sponsorship or employment prohibit you from disclosing the data. Therefore, check your contractual obligations before revealing your data.

3.3.1 Sharing your research findings at conferences

At scientific conferences, researchers often present their most current research findings, which are not yet published as complete research papers, as either poster or oral presentations (see Chapters 13 and 14). Attendance at conferences will give you the opportunity to present your preliminary data in either poster or oral format and expose you to the most current thinking in the field. In addition, you will have the opportunity to meet and engage in discussion with other scientists working in the same field as you or a related field (see Chapter 15). Therefore, it is extremely important that you attend as many conferences as you are able to during your research programme.

At conferences, most bioscientists will discuss their research to the extent that other scientists have a general overview of the work they are conducting. However, they may avoid disclosing detailed information about their work or their ideas. The commonly cited reason for this is that the data could be 'scooped'; that is, another research group could use your ideas and take the credit by publishing the work before you. However, the advantage of sharing information is that you will benefit from the input of other scientists. When at conferences, you will need to make a careful judgement about the level of information to disclose at the early stages of your unpublished work, particularly if you are discussing your work with a competing research group. Use the considerations outlined in section 3.3 as a starting point to guide you, and discuss with your supervisor(s) for additional advice.

3.3.2 **Publishing your research findings**

As your work progresses, you will build up sufficient data to publish a research paper in a peer-reviewed journal. At this stage, you will communicate the precise details of your experimental methodology and results so that independent scientists working in the same field as you can repeat the work and confirm or refute it, or at least evaluate the quality of your findings (see Chapter 6). The publication process is described in Chapter 4 and guidelines on how to write research papers are presented in Chapter 9.

As a research scientist you have a responsibility to report your results and share the associated data sets with the wider scientific community in a timely manner. This exchange of information facilitates scientific discovery and is therefore of benefit to both the researchers and the public (Suber, 2007). There are two main mechanisms for disseminating your findings in a way which will make them more widely and openly accessible. These are open access publishing, and self-archiving your papers and data sets.

Open access journals

In an open access journal, the cost of publication is borne by the author(s) (commonly their funders or institution) and the article is available to the reader free of charge. This differs from the traditional subscription-based journal, in which the costs of publication are met by the publisher and the article is restricted to subscribers only. The two main open access publishers are Biomed Central and the Public Library of Science (PLoS). However, more and more subscription-based publishers are now giving authors the option to publish their article as open access or through the regular subscription model within the **same** journal (Rinaldi, 2008). This hybrid model is operated by publishers such as Oxford University Press, Elsevier, Blackwell/Wiley, Nature Publishing Group, and Springer. For a list of open access journals, see the Directory of Open Access Journals (http://www.doaj.org).

Open access publishing has a significant advantage over traditional methods of publishing in that the paper is immediately accessible to the public and not restricted to those who pay subscription fees. Another advantage is that copyright is retained by the author and the work is covered by a Creative Commons type licence (see section 3.6). These licences allow copyright holders to specify conditions under which their work may be used. For example, work covered by the attribution licence allows the public to copy, distribute, and display the work as long as the author(s) are acknowledged appropriately. This removes some of the restrictions imposed by copyright and therefore the research findings can be more easily used by the research community.

Free-to-access archives

You can deposit a electronic copy of your published paper into a free-to-access archive. Archives can contain preprints of research papers, postprints, or both. A preprint is a version of a research paper which has not been peer reviewed or published; a postprint is a paper that has been peer reviewed and published (see Chapter 4, section 4.2). An example of a preprint archive is Nature Precedings, and examples of postprint archives are PubMed Central (PMC) and UK PubMed Central (UKPMC). The latter two archives provide free access to the full content of peer-reviewed published research articles. If you decide to deposit a copy of your published paper in an open access archive, then check the conditions of copyright (see section 3.6). Some journals will allow you to archive your paper and some may not.

Depositing data in a databank

In addition to archiving your published paper you are often required to archive the final data sets and materials associated with your paper into a public databank. Examples include depositing cell lines into a repository such as the American Type Culture Collection, DNA sequences into GenBank or EMBL, and ecology and biodiversity data into Knowledge Network for Biocomplexity.

Motivations for sharing papers and data sets

There are a number of factors which may influence your decision to publish in open access journals or deposit your research data in free-to-access archives and databanks.

- You may be motivated to share your papers and data sets as a consequence of your personal convictions that data should be shared freely amongst scientists.
- You may be motivated to share your papers and data sets as it will make your work more widely available and therefore potentially increase the number of times your work is cited by other authors. That an open access article is more likely to be cited than an article not published or archived under open access has been confirmed by some studies but challenged by others (Craig *et al.*, 2007).
- It may be a requirement of your funding organization. Some funding organizations such as the Wellcome Trust, the US National Institutes of Health, and the European Research Council require that papers published in peer-reviewed journals that are a product of their funds are deposited in a designated archive within a designated period of time after the paper is published.
- It may be a requirement of the journal in which you publish that the data associated with the publication are deposited in a suitable databank. Many journals expect authors to archive the data associated with their publication and will not accept the article for publication until an entry code to a suitable databank has been provided.

3.3.3 Interpreting and reporting your research findings

When discussing or reporting your research data you must describe and interpret your data accurately and completely. This means you should not invent data, nor should you hide part of your data or distort your findings in order to show or strengthen a particular conclusion. If you fabricate data or misrepresent your work, then other people will waste valuable time attempting to validate your findings. If you are found out, then it could cost you your professional reputation or your job, and could have legal consequences. For a brief summary of reported cases and consequences of data falsification and fabrication refer to Odling-Smee *et al.* (2007).

To avoid unethical interpreting and reporting of data:

- Maintain detailed and accurate experimental notebooks (see Chapter 2, section 2.1). This will help you write up your work accurately.
- When reporting your work, include all the findings that are relevant to your research questions including those that may not support your original hypothesis. You should not withhold data or alter your data in order to show or strengthen a particular conclusion or based on a 'gut feeling' they are inaccurate. Examples of practices that can distort your results and hence your conclusions include:
  - Excluding outliers in data sets.
  - Using different statistical tests until you find a significant result that supports your hypothesis and then reporting the significant statistical analysis only.
  - Failing to report results which do not support your hypothesis.
  - Designing graphs in such a way that the trends are visually distorted.
  - Enhancing some parts of a gel (or other image) relative to others, using image processing software.

For a comprehensive list of the types of unethical practices that can distort the results and hence the conclusions and therefore must be avoided, refer to Al-Marzouki *et al.* (2005) and Martinson *et al.* (2005).

- It is acceptable, however, to report certain data in full and summarize other findings. This is called **data filtering** and, if you use it, then you will need to justify fully your reasons for doing so (Council of Biology Editors, 1990).
- You must disclose full details of how your data were generated and how they were analysed. In particular, if any images (such as gels, immunoblots, immunohistochemistry) have been electronically processed in any way, then details of how the images were processed should be clearly described and the originals without any manipulations retained. When using graphics editing software you must keep the processing to a minimum. You should not in any way 'manipulate your image to enhance, clarify or conform to an expected result' (Nature, 2004, 2006). This is termed 'data beautification' and is a form of misrepresentation. For an excellent review of the types of image manipulation which are unacceptable, consult Rossner and Yamada (2004). This article provides some specific examples of inappropriate practices based on real cases of digital manipulation discovered through inspection of digital images (micrographs and electrophoretic gels and blots) in a sample of papers submitted for publication in a journal.
- You must acknowledge fully and accurately any work of other authors that you use in your own work. To not acknowledge or to acknowledge incompletely constitutes plagiarism. This is discussed in detail below.

3.4 Understanding plagiarism

A good piece of academic work shows evidence of original thinking and uses the work of others in order to place **your work** in the context of the wider published literature and to support **your own** analysis, proposals, and interpretations. When using the work of others you must use it appropriately. That is, you must credit the original author(s) by referencing the source in your work. If you do not differentiate clearly between your own work and the work of others, you may be guilty of plagiarism. The punishments for plagiarism can be very severe. If you plagiarize an assignment, then it could result in the assignment being awarded a mark of zero, or you could fail the course or be expelled from the university. If you plagiarize a piece of work and publish it (or attempt to publish it), then you could be tried in a court of law.

Plagiarism can be defined as 'Presenting someone else's work as your own. Work means any intellectual output and typically includes text, data, sound or performance, (University of Leeds, 2005).

Plagiarism can take a variety of forms, including the following:

- Copying verbatim another person's work without crediting the original author.
- Careless paraphrasing of another person's work so that the original text is changed slightly but still bears close resemblance to the original text, with or without acknowledging the original author.
- Insufficient referencing so that it is difficult to differentiate clearly between work that is your own and the work of others. This could include acknowledging only part of what is used from a published source or failing to cite the source in the text at the point you refer to it; for example, by placing references at the end or beginning of paragraphs only. Insufficient referencing also includes failing to reference information obtained through personal communications such as e-mail correspondence, telephone conversations, and face-to-face discussions.
- Reusing sections of work that you have previously authored either in another assignment or in a previous publication could also be considered plagiarism (depending on the type and extent of reuse) and is termed self-plagiarism.
- Plagiarism is not limited to copying a tangible output such as text, data, sound, or performance, but also copying someone else's ideas without proper attribution.

3.4.1 Avoiding plagiarism

There are many resources you can consult to help you avoid plagiarism. These include advice from your tutors, information provided by your university, and publications on plagiarism such as the book by Neville (2007). Section 3.7 includes examples of plagiarized text and an example of how to use the work of others appropriately in your work. These examples should help you to understand how the work of others can be used in your work in an academically acceptable way.

The strategies you can use to avoid plagiarism are described briefly below.

Referencing completely and accurately

Whenever you use another person's material in your work, then credit the original author by referencing completely and accurately. Complete and accurate referencing means:

- Citing each source within the text at the point you refer to it.
- Listing fully all the sources that you have cited in the text at the end of your writing in the form of a reference list. For comprehensive guidelines on how to reference see Chapter 1, section 1.2.

You should reference whenever you use another person's facts, data, illustrations, theories, opinions, interpretations, or ideas that are not considered common knowledge. Common knowledge is information which many people have observed and is therefore widely known by the specialist community. For example, the statement

> DNA is a polymer of deoxyribonucleotide units. Each nucleotide consists of a nitrogenous base, a sugar (deoxyribose), and one or more phosphate groups

is considered common knowledge and therefore does not require a citation. However, the following statement:

> Vertebrate DNA can be methylated at the cytosine of the dinucleotide CpG and this may act as a stabilizer for inactive chromatin during normal development (Bird, 2002)

requires a citation as this information is not widely known. If you are unclear whether something is common knowledge or not, then you should reference to avoid a possible charge of plagiarism.

You should acknowledge not only published material but also information that is derived from personal communications such as e-mail correspondence, telephone conversations, and face-to-face discussions. Personal communications can include unpublished data obtained from your research group members or members of another research group. It could also include discussions of your work with other individuals that lead you to a new insight into your work. In both cases, you should seek permission before using the material in your work and cite the source appropriately (see Chapter 1, section 1.2).

Quoting the work of others appropriately

If you decide to use verbatim a passage of text from a published source in your work then you will need to enclose the borrowed text in quotation marks **and** credit the original author. If you use quotations, you should do so minimally and only when you are going to discuss a statement in the context of your own work. On the whole, the use of quotations in scientific writing is not common practice and should be used sparingly or avoided altogether.

Paraphrasing the work of others appropriately

Paraphrasing means rewriting another person's work in your own words. When you paraphrase, make sure you do it completely so that the sentence structure and the words are distinctly your own. Do not simply substitute a few words and/or change the order of a few words in the original sentence, as this will be considered plagiarism. When paraphrasing, quotation marks are not required but you will need to cite the source of information. When you paraphrase, take care to convey the meaning of the original author's work accurately. This is only possible if you understand clearly the content you are reading. You should therefore understand the material you read, write it in your own words, and then check your work against the original source for accuracy of meaning.

Using published illustrations in your work

When using illustrations (such as diagrams and figures) in your work which is for **assessment purposes**, you may either use the published illustration intact or you may adapt it. In both instances, you will need to acknowledge the source from which the illustration was taken or adapted. If you intend to use an illustration in a piece of work intended **for publication** then you will need to obtain permission from the copyright owner and cite the source in your work (see section 3.6).

When using an illustration, always consider if it will add to the understanding of your work. Sometimes illustrations can be very useful in helping to visualize difficult models and concepts. At other times, illustrations may not be necessary as the information can be summarized in a few sentences. It is poor academic practice to include many illustrations that are taken from various sources, even with appropriate acknowledgement, as it will show a lack of original thinking.

Reusing your own work

In some cases an individual may reuse sections of work that they have previously authored, either in another assignment or in a previous publication. If you are writing for assessment purposes, then you will not be able to submit the same piece of work for assessment twice, either in part or whole, unless you have been granted permission to do so. Most universities require that each piece of work submitted for assessment is original and that you are not awarded credit for the same piece of work twice. It is possible that a situation may arise where you are required to submit a piece of work on a topic you have previously authored for assessment purposes (that is, a mark that contributes to the classification of a degree). This situation may arise, for example, when you have already submitted for assessment an initial proposal of your research work which includes a substantial amount of literature review and then you write a dissertation on the same project which also contains a review of the literature. In this case, it is advisable not to reuse the material in its original format but instead to transform it in a way that will make it an original piece of writing. You could do this by extending on a theoretical explanation or discussing a wider number of relevant studies in your work. Take advice from your tutors if you are unclear about the reuse of material you have previously authored.

If you are publishing a piece of work, most journals allow a **small degree** of overlap in a limited number of instances. Examples of acceptable reuse could be to reuse a limited part of an introduction from an earlier paper (Giles, 2005) or to reuse a figure you have previously authored. In both cases, remember that that the publisher, not you, may own the copyright to your earlier paper and therefore you may have to seek permission before using it (see section 3.6). If you recycle a substantial amount of your work, then the paper may not be accepted for publication (see section 3.5.1).

3.5 Duplicate submissions, duplicate publications, conflicts of interest, and authorship issues

This section focuses on key ethical practices that relate to the publication of research data. These are duplicate submissions and publications, conflicts of interest, authorship, and acknowledgements.

3.5.1 Duplicate submissions and duplicate publications

When you submit a research paper to a journal for publication, you must not submit the manuscript simultaneously to two (or more) different journals. However, it is completely acceptable to resubmit to a different journal if your paper is rejected. In addition, the manuscript must not be a duplicate publication. A duplicate publication is one where the full paper (or considerable parts of it) has been published previously. If your research findings have been previously published in the form of a meeting abstract this does not constitute a prior publication and therefore does not prevent you from submitting your complete findings in the form of a research paper. Uploading your manuscript on to a recognized preprint server such as Nature Precedings for a review by other scientists in the field prior to formal submission to a journal is acceptable practice for some (but not all journals). You should therefore always consult the policy of the journal you intend to publish in before you post on a preprint server. It is also acceptable to include a previously published illustration in your manuscript as long as permission has been obtained from the copyright holder (see section 3.6) and the source of information is clearly referenced.

It is not acceptable to recycle large parts of your introductory text, although a small degree of overlap may be acceptable (see section 3.4.1). It is also unacceptable to divide a study which can be reported in a single paper into smaller studies and then publish them as multiple papers. This is called 'salami slicing' and wastes valuable publishing space.

3.5.2 Conflicts of interest

Conflicts of interests are circumstances which could potentially lead to (or could be perceived as leading to) a bias in the way you, as an author, present your research findings. These conflicts of interests could be financial (for example, the source of your research funding), personal (for example, relationship with the journal editorial board), or professional (for example, membership of a government advisory board). At the time you submit your manuscript you will be asked to disclose any conflicts of interest to the editor of the journal. Your disclosure will allow the editor to review your manuscript with knowledge of all the potential conflicts of interest and ensure that the findings and their interpretations have not been influenced inappropriately. When the paper is published the conflicts of interest are disclosed to the public. This usually appears as a statement at the end of the paper that may read, 'no conflicts of interest declared' or as a statement that lists the conflicts of interests.

Disclosing conflicts of interest and then evaluating which conflicts are acceptable is important so that the public's confidence in the integrity of scientific research is not eroded. A number of well-publicized cases where conflicts of interest have led scientists to misrepresent their findings—such as downplaying the risks of passive smoking whilst receiving funding from tobacco companies (Neroth, 2005)—have led scientists to propose ways in which conflicts of interest can be managed. These include disclosure of conflicts of interest policies, banning stock ownership (Kaiser, 2005), and more robust mechanisms for evaluating how acceptable specific conflicts of interest are (Hurst and Mauron, 2008). Remember, it is not necessarily wrong that a conflict of interest exists but it is wrong that it should lead you to report your research dishonestly.

3.5.3 Authorship

People almost always conduct research work in teams which include multiple people working on different aspects of the same project. These teams may be composed of people from the same research group or people from different research groups, different disciplines, different institutions, and even different countries. Working in teams can raise critical ethical questions around authorship of research papers. The three main questions are:

1. Who will write the paper?
2. Who will be listed as a co-author?
3. In which order should each qualifying author be listed on the manuscript?

Who will write the paper?

If you are the research student or postdoctoral research fellow who has conducted a significant part of the study and interpreted the data, then typically you will write the first draft of the paper. This first draft will then be revised by your supervisor and other co-authors before it is ready for submission to a journal. In some research groups, the supervisor may write the first draft with input from the student. Writing research papers is an important part of your training and you should as far as possible write the first draft yourself.

Who qualifies for authorship?

Currently there are no agreed criteria for identifying who should be listed as an author of a particular manuscript. Different authors, different disciplines, and different institutions practise different policies when defining authorship (Marco and Schmidt, 2004). However, standard guidelines provided by the International Committee of Medical Journal Editors (ICMJE, 2006) state that an author is one who has made **substantial** contributions to:

- the conception and design of the study, collection of data or analysis, and interpretation of the data **and**
- drafting or revising the manuscript for important content **and**
- approving the final version for publication.

This definition has been adopted by a number of journals and you could use this recommendation as a benchmark for assessing authorship. The journal you intend to publish in may also provide guidance on authorship, which may be the ones recommended by the ICMJE or similar to them.

If you believe that you have made a significant contribution to a particular project and deserve to be credited as a co-author, but have been omitted from the list of authors, then you should be able to obtain advice from the support structure that your department has in place for supporting students. This could involve talking to an advisor (or other identified person) allocated to mentor you during your programme of study, or your departmental head.

By identifying yourself as an author of a manuscript you take credit for the published work and you take public responsibility for its integrity. It is therefore essential that your work is conducted and interpreted honestly so that you do not damage your reputation as a scientist nor damage the reputation of your colleagues who are listed as co-authors. At the time you submit your manuscript, you may be expected to sign an authorship form confirming that you have contributed to the study in a meaningful way and that the work has been conducted and reported in an honest manner. Some journals also request that descriptions of the type and level of contribution made by each individual author are also submitted to the journal at the same time as the manuscript.

Order in which authors should be listed

It is common for the first author to be the principal researcher who has conducted a significant part of the study, interpreted the data, and drafted the article, and the last author is often the one who has supervised and overseen the study. The remaining co-authors are placed in descending order (after the first author) according to the level of contribution made. However, this practice may vary between research groups and institutions and you should therefore consult the institution at which you work for local practices. Determining the order in which each author should be listed on a manuscript can cause significant conflict between authors. Again, if you disagree with the order of authorship, you should attempt to discuss this with your supervisor and, if this is not satisfactory, then with your mentor or departmental head.

3.5.4 Acknowledgements

Each paper has an acknowledgements section which consists of a short paragraph in which the authors acknowledge the assistance provided by any people or institutions that are not listed as co-authors.

From the recommendations made by the ICMJE (section 3.5.3) we can see that to qualify for authorship you must make a **substantial** contribution to either the concept and design of the study, or to data collection, or to data analysis and interpretation, **and** contributed to the writing and revising, **and** approved the manuscript. Anyone who has contributed to the work but does not meet these authorship recommendations should be considered for an acknowledgment. Examples of contributions which merit acknowledgement are:

- Source of funding that supported your research.
- Source of technical assistance.
- A researcher or research group who supplies you with material such as a starting plasmid construct or an antibody.
- A colleague who reviews drafts of your manuscript and makes suggestions for improvement.

3.6 Using copyright material in your work

When using material from published sources, you must use it in a way that does not violate copyright laws. This section summarizes the basics of copyright and how to use copyright-protected material in your work appropriately. The information supplied here is based on UK copyright laws as set out in the Copyright, Designs and Patents Act 1988 (the '**Act**') (available from the Office of Public Sector Information, http://www.opsi.gov.uk/). Copyright laws vary between different countries and therefore you will need to consult the office in charge of intellectual property in your country (if not the UK) for regulations governing the use of copyright material. You may also want to consult the Council of Science Editors manual (CSE, 2006) which summarizes copyright law for the USA and identifies differences between the copyright laws of Australia, Canada, New Zealand, and the UK.

3.6.1 What is copyright?

Copyright grants the copyright owner the exclusive right to distribute, reproduce, and adapt their work (and to authorize others to do the same) for a certain number of years. Work that is eligible for copyright includes original literary works (such as books, journals, manuscripts, and computer programs), dramatic, musical, and artistic works, sound recordings, films and broadcasts, published editions of works, and databases. The number of years that copyright lasts varies for different types of original works. This information is listed in the *Copyright basic facts* booklet available from the UK Intellectual Property Office (2007) to which you should refer. In the UK, copyright protection begins immediately once the original work is recorded or written down in a fixed and permanent form. It is not a legal requirement that the work is officially registered with an intellectual property office or that it carries the copyright symbol © for the work to be considered copyright protected.

3.6.2 Who owns the copyright?

In most cases the creator of the work is the first copyright owner. However, two exceptions to this rule under which the creator of the work is not the copyright owner (but this is not an exhaustive list) are:

- **When the work is created by an employee during the course of his/her employment with another person or organization**. In this instance the employer and not the employee holds the copyright to the work produced, subject to any agreement to the contrary. An example of this is when research is conducted at a particular institution. In such circumstances the owner of the research is likely to be the institution and not the researcher. In some cases, ownership may be shared between the institution and a third party (such as a funding body) (see section 3.2).
- **When copyright is transferred to another person or organization**. For example, authors of academic work usually transfer copyright to the publisher so that the publisher holds the copyright and not the author of the work.

3.6.3 How to use copyright material appropriately in your work

This section provides some general guidelines on using copyright material in your work and is based on UK copyright law.

Can I photocopy journal articles for my own research or private use?

Fair dealing is a permitted act under copyright law that allows you to copy limited amounts of copyright material for your own private study or non-commercial research without requesting permission from the copyright owner and without violating UK copyright laws. It is not possible to say exactly what amount of copying legally constitutes 'fair dealing' as the Act does not provide specific guidance on this question. This means that it will ultimately come down to what is reasonable in the circumstances. However, making a single photocopy of a journal article from a hard copy journal issue for your own research or private use is generally considered to be fair dealing. You may not, however, make multiple photocopies of the article.

Can I save and print electronic copies of journal articles for my own research or private use?

As stated above, it is difficult to say accurately exactly what constitutes fair dealing. However, guidance published by the Joint Information Systems Committee (JISC) and the Publishers Association (*Guidelines for fair dealing in an electronic environment,* 1998) indicates that you may download and save a single copy of an electronic journal article from a journal issue for non-commercial research or your own private use and this will constitute fair dealing. You may not, however, download or save a complete journal issue. You may also print out a single hard copy of an electronic journal article from a journal issue. If it is an electronic book, you may save and copy a single chapter from a book and you may save and print out a single copy of a web page. You may not, however, make further (paper or electronic) copies from that original saved or printed copy.

Can I use copyright material for assessment purposes?

It is generally considered acceptable to use copyright material (for example, illustrations or portions of text or sound or performance) in work which is intended for non-commercial assessment, such as, reports, theses, dissertations, and web-based assessments. However, you must use it appropriately in your work so that charges of plagiarism are not made. This means genuinely discussing the material in the context of your own work and referencing the material fully and accurately (see section 3.4) so that there is no potential for the reader to assume incorrectly that the copyrighted material is your original work.

How do I use copyright material in work that is intended for publication?

If you want to use copyright material in work which is intended for publication then, unless otherwise stated, you will need to obtain permission from the copyright holder **and** acknowledge the original source completely and accurately in your work. Permission to use the work may be obtained by putting your request **in writing** to the copyright owner or by writing directly to an organization which represents copyright owners (e.g. the Copyright Licensing Agency). When writing your letter to request permission to reuse the work, state clearly (1) the **precise** material you wish to use, (2) the **precise** use you wish to make of the work, and (3) the **type of media** in which the work will be printed, that is print or electronic format. Make sure that you obtain permission in writing prior to any publication and keep this for your records. If permission is granted, you may be required to pay a fee for the initial use and any reuse of the material.

However, there are instances in which you may be able to use the material for publication purposes without requesting permission. These are:

- **Fair dealing** allows you to use limited amounts of text for criticism or review without requesting permission of the copyright owner, provided you fully acknowledge the copyrighted text and it has already been made available to the public. The exact amounts are not specified by the Act, but the guidelines provided by the Society of Authors (1965) state that using a single extract of 400 words, or a series of extracts to a total of 800 words (of which none exceeds 300 words), in the context of review or criticism is likely to be considered fair dealing.
- If the work is covered by a UK **Creative Commons licence** (Creative Commons, 2008) or a similar open-licensing agreement, then you may be able to use the work without requesting permission. Creative Commons licences allow copyright holders to specify conditions under which their work may be used. For example, work covered by the Attribution (by) License (which is the most accommodating of the Creative Commons licences) allows you to distribute, build upon, and display copyright work, even commercially, as long as you acknowledge the original source appropriately. Some Creative Commons licences will also allow you to adapt the work. Work covered under the Creative Commons licence carries a Creative Commons statement. You should check the conditions of the licence before using the material in your work, particularly as some of the licences do not permit commercial use of the work. If you want to use the material for purposes and/or in ways other than that permitted by the licence, then you will need to contact the copyright holder to obtain written permission prior to using the material in such a way.

3.7 Examples of plagiarized work and an example of how to use the work of others appropriately

A good piece of academic writing should show evidence of original thinking and use references to support **your own** analysis, proposals, and interpretations. This section illustrates two examples of inappropriate use of material which constitute plagiarism. The first is an example of cutting and pasting from an original source and not acknowledging where the information is taken from. The second example is a more subtle case of plagiarism involving careless paraphrasing. These two examples of plagiarism are followed by a third example which illustrates how you can use the work of others in a way which is academically acceptable.

Let us consider a situation where a student is asked to write a review on the following topic 'The reproductive cloning of mules'. The original extract of the plagiarized work is included below. The source is a news article published in the journal *Science*.

Original extract from Holden, C. (2003). First cloned mule races to finish line. *Science*, 300, 1354 (reprinted with permission from AAAS)

The first equine has joined bovines, ovines, felines, rabbits, rodents, and porkers in the ranks of the cloned. On 5 May a mule named Idaho Gem was born after a normal 346-day gestation in the womb of a mare, researchers report online in *Science* this week (http://www.sciencemag.org/cgi/content/abstract/1086743). That makes him not only the first member of the horse family but also the first sterile animal to be cloned. Mules, sired by donkeys and borne by horses, are incapable of reproduction. But a team at the University of Idaho in Moscow headed by Gordon Woods has now shown that a mule cell nucleus, despite its odd chromosome number, can cut the mustard.

Idaho Gem is a sibling of a world-champion racing mule named Taz. The scientists didn't want to clone from an adult animal because they 'wanted to take the aging component out of the equation,' Woods says. Some researchers suspect that the first clone, Dolly the sheep, aged prematurely because her DNA was derived from an adult cell. So the team rebred Taz's parents, took a somatic cell from the 45-day-old fetus, and fused it with an enucleated horse oocyte that they then inserted into a mare.

Equines have proved difficult to clone; horse oocytes don't mature well in a dish and it's hard to get embryonic cells to divide. Woods's group found that calcium levels inside equine red blood cells are low compared with those from cows, leading the researchers to suspect that low calcium levels could be inhibiting cloned equine embryos' growth. They jacked up the calcium in the cultures and got some embryos to thrive.

3.7.1 Examples of plagiarized work

Example 1

Reproductive cloning of mules

The first equine has joined bovines, ovines, felines, rabbits, rodents, and pigs amongst cloned species. On 5 May a mule was born after a 346-day gestation in the womb, researchers report online in *Science* this week (http://www.sciencemag.org/cgi/content/abstract/1086743). According to Science, that makes him not only the first member of the horse family but also the first sterile animal to be cloned. Mules, fathered by donkeys and borne by horses, are sterile. A group at the University of Idaho in Moscow has now shown that a mule cell nucleus, despite its odd chromosome number, can function.

Idaho Gem is a sibling of Taz, a world-champion racing mule. The scientists did not want to clone from an adult animal because they preferred to remove the aging component from the equation. It is suspected that the first clone, Dolly the sheep, aged prematurely because her DNA was derived from an adult cell. In this case the team rebred Taz's parents and used a somatic cell from a 45-day-old fetus to fuse with an enucleated horse oocyte which was then inserted into a mare.

Members of the horse family have proved difficult to clone; horse oocytes do not mature well in culture and their embryonic cells fail to divide. New data showed that calcium levels inside equine red blood cells are low compared with those from cows, leading the researchers to suspect that low calcium levels could be inhibiting cloned equine embryos' growth. Increasing the calcium in the cultures promoted the survival of some embryos.

If you read through the above example and compare it with the original source, you will see that it is a clear case of plagiarism because:

- The writer has copied large amounts of text directly from the original source and merely added a few words between sentences.
- The writer has not acknowledged the source from where the information was taken.

If the writer had placed the copied text in quotation marks and cited the original source of information then technically this would not be considered plagiarized work. However, the quality of work would be considered extremely poor as there is no evidence that the writer has understood the material or used it in an original way (e.g. to support the writers' own interpretation or analysis). Remember, if you use quotations, you should do so minimally and only when you are going to discuss a statement in the context of your own work.

Example 2

Reproductive cloning of mules

For the first time a member of the horse family has been cloned. This week *Science* reported the cloning of a mule after a normal pregnancy (http://www.sciencemag.org/cgi/content/abstract/1086743). Mules, which are produced by crossing donkeys and horses, are sterile, making 'Idaho Gem' the first cloned sterile animal.

To avoid using cells from a mature animal the parents of a prize-winning racing mule, Taz, were re-mated to provide the starting material. This was done because of fears that the original cloned sheep, Dolly, had aged too quickly because her DNA was derived from an adult cell. A somatic cell from a six-week old fetus produced by Taz's parents was fused with a horse egg cell, the nucleus of which had previously been destroyed. The fused cell was subsequently placed into a surrogate mother horse.

Horses have previously been refractory to cloning techniques because of the difficulty of growing their cells in culture. New research compared calcium levels in red blood cells from horses and cows and found that increasing the levels to those found in cows resulted in viable horse embryo cultures.

In this example the writer has made some attempt to paraphrase from the original article. However, this work is still considered plagiarism because:

- The content and structure are still similar to the original source. The writer has merely substituted or omitted a few words or changed the order of the words in the original article. This type of careless paraphrasing is unacceptable as the structure and content of the original work are still retained.
- The writer has not acknowledged the source from which the information was taken.

If the writer had paraphrased completely, that is rewritten the work in her own words, and cited the source of information, then this would not be considered plagiarized work. However, the quality of work, although better than the first example, would still be weak as it does not show evidence of wide reading or independent analysis.

3.7.2 Example of acceptable use of published material

The following example shows how the work of others can be used appropriately in your work.

Example 3

Reproductive cloning of mules

Somatic cell nuclear transfer, or cloning, describes the production of genetically identical animals by inserting a cell nucleus from the animal to be cloned into an enucleated oocyte from another individual. The best known example of using such nuclear transfer technology to produce a cloned animal was Dolly the sheep (Campbell *et al.*, 1996). Cloning by somatic cell nuclear transfer has now been successfully demonstrated in many animal species, including mice (Wakayama *et al.*, 1998), cattle (Kato *et al.*, 1998) pigs (Polejaeva *et al.*, 2000), goats (Zou *et al.*, 2001), cats (Shin *et al.*, 2002) and rabbits (Chesne *et al.*, 2002). However, so far, it has been impossible to apply current cloning techniques to some other species of interest, including horses. In a recent issue of *Science*, modifications to the cloning methodology have been reported, which have made it possible to clone the first mule (Woods *et al.*, 2003; Holden, 2003). This research might make it possible to extend cloning to other members of the horse family.

The failure of standard cloning techniques in some species could be due to many factors. It is possible, for example, that species specific differences in the timing of onset of nuclear gene expression in embryos might influence cloning efficiency under standard conditions (Renard *et al.*, 2002). However, in the case of horse cloning, it appears possible that the primary cause of failure has simply been unfavourable culture conditions. In this study by Woods *et al.*, (2003), a comparison of calcium levels in the red blood cells of cows and horses revealed that the calcium levels were lower in horse cells. Increasing the calcium levels in horse embryo cultures resulted in increased viability and contributed to the production of the first cloned mule. It now remains to be seen whether this modification will lead to the production of cloned horses as well as mules, which are the result of crossing a donkey and a horse. This finding also raises the possibility that optimization of culture conditions might facilitate cloning in other species which have previously been difficult to clone.

It has been suggested that Dolly showed signs of premature ageing and this has been attributed to the use of cultured adult cells to provide the donor nucleus (Shiels *et al.*, 1999). Although this is far from certain, Woods *et al.* (2003) in their current study used fetal tissue, derived from re-mating the parents of a champion racing mule called Taz. It remains to be seen whether the theoretical benefit derived from the use of fetal tissue nuclei outweighs the obvious disadvantage of using sibling material to clone a racing mule. However, the resulting clone will not be a clone of Taz.

Reference list

Campbell, K.H.S., McWhir, J., Ritchie, W.A., Wilmut. I. (1996). Sheep cloned by nuclear transfer from a cultured cell line. *Nature,* **380**, 64–66.

Chesne, P., Adenot, P.G., Viglietta, C., Baratte, M., Boulanger, L., Renard, J.P. (2002). Cloned rabbits produced by nuclear transfer from adult somatic cells. *Nat. Biotechnol.*, **20**, 366–369.

Holden, C. (2003). First cloned mule races to finish line. *Science,* **301**, 1354.

Kato, Y., Tani, T., Sotomaru, Y., Kurokawa, K., Kato, J., Doguchi, H., Yasue, H., Tsunoda, Y. (1998), Eight calves cloned from somatic cells of a single adult. *Science,* **282**, 2095–2098.

Polejaeva, I.A., Chen, S.H., Vaught, T.D., Page, R.L., Mullins, J., Ball, S., Dai, Y., Boano, J., Walker, S., Ayares, D., Colman, A., Campbell, K.H.S. (2000). Cloned pigs produced by nuclear transfer from adult somatic cells. *Nature,* **407**, 505–509.

Renard, J.P., Zhou, Q., LeBourhis, D., Chavatte-Palmer, P., Hue, I., Heyman, Y., Vignon, X. (2002). Nuclear transfer technologies: between successes and doubts. *Theriogenology,* **66**, 6–13.

Shiels, P.G., Kind, A.J., Campbell, K.H.S., Waddington, D., Wilmut, I., Colman, A., Schnieke, A.E. (1999). Analysis of telemere lengths in cloned sheep. *Nature,* **399**, 316.

Shin, T., Kraemer, D., Pryor, J., Liu, L., Rugila, J., Howe, L., Buck, S., Murphy, K., Lyons, L., Westhusin, M. (2002). A cat cloned by nuclear transplantation. *Nature,* **415,** 859.

Wakayama, T., Perry, A.C., Zuccotti, M., Johnson, K.R., Yanagimachi, R. (1998). Full-term development of mice from enucleated oocytes injected with cumulus cell nuclei. *Nature,* **394**, 369–374.

Woods, G.L., White, K.L., Vanderwall, D.K., Li, G.P., Aston, K.I., Bunch, T.D., Meerdo, L.N., Pate, B.J. (2003). A mule cloned from fetal cells by nuclear transfer. *Science,* **301**, 1063.

Zou, X., Chen, Y., Wang, Y., Luo, J., Zhang, Q., Zhang, X., Yang, Y., Ju, H., Shen, Y., Lao, W., Xu, S., Du, M. (2001). Production of cloned goats from enucleated oocytes injected with cumulus cell nuclei or fused with cumulus cells. *Cloning,* **3**, 31–37.

This work is **not** considered plagiarism because:

- The writer has assimilated material from various sources and produced an original piece of work so that the content and structure are distinctly her own.
- All the sources are accurately and completely cited in the body of the text and then listed at the end of the review in the form of a reference list.

This example shows convincingly that the student had understood the subject matter and used published material from a variety of sources to support her own analysis of reproductive cloning in mules.

To summarize, you must be completely ethical in the way you use the work of others to communicate your own work. You should not intentionally attempt to pass off the work of others as your own, nor should you unintentionally plagiarize as a result of careless paraphrasing or insufficient referencing.

Many universities are now using plagiarism detection software such as Turnitin to detect incidents of plagiarism by students. Turnitin highlights similarities between text uploaded on to the software and the work of other students simultaneously or previously uploaded into the system. Turnitin also matches uploaded text with electronic material available on the web, including commercial databases of journal articles—and hence plagiarism is easily detectable. However, you should avoid plagiarizing, not because it is easily detectable, but because of a commitment to academic integrity—that is, the fair and honest use of the work of others.

References

Al-Marzouki, S., Roberts, I., Marshall, T., Evans, S. (2005). The effect of scientific misconduct on the results of clinical trials: a Delphi survey. *Contemp. Clin. Trials*, **26**, 331–337.

Benos, D.J., Fabres, J., Farmer, J., Gutierrez, J.P., Hennessy, K., Kosek, D., Lee, J.M., Olteana, D., Russell, T., Shaikh F., Wang, K. (2005). Ethics and scientific publication. *Adv. Physiol. Educ.*, **29**, 59–74.

Bird, A. (2002). DNA methylation patterns and epigenetic memory. *Genes Dev.*, **16**, 6–21.

Council of Biology Editors (1990). *Ethics and policy in scientific publication.* The Council, Bethesda, MD.

CSE (2006). *Scientific style and format: the CSE manual for authors, editors, and publishers*, 7th edn. Council of Science Editors in cooperation with Rockefeller University Press, Reston, VA.

Craig, I.D., Plume, A. M., McVeigh, M.E., Pringle, J., Amin, M. (2007). The effect of use and access on citations. *J. Infometrics*, **1**, 239–248.

Creative Commons (2008). *License your work*. [online]. Available at: http://creativecommons.org/license/ (last accessed 7 April 2008).

ICMJE (2006). *Uniform requirements for manuscripts submitted to biomedical journals: writing and editing for biomedical publication* [online]. International Committee of Medical Journal Editors. Available at: http://www.icmje.org (last accessed 10 September 2007).

Giles, J. (2005). Where to draw the line? *Nature*, **435**, 258–259. doi: 10.1038/435258a.

Hurst, S.A. and Mauron, A. (2008). A question of method. *EMBO Rep.*, **9**(2), 119–123.

JISC/PA (1998). Joint Information Systems Committee and the Publishers Association. *Guidelines for fair dealing in an electronic environment* [online]. Available at: http://www.ukoln.ac.uk/services/elib/papers/pa/fair/intro.html (last accessed 7 April 2008).

Kaiser, J. (2005). NIH chief clamps down on consulting and stock ownership. *Science*, **307**, 824–825.

Ledford, H. (2008). With all good intentions. *Nature*, **452**, 682–684.

Marco, C.A. and Schmidt, T.A. (2004). Who wrote this paper? Basics of authorship and ethical issues. *Acad. Emerg. Med.*, **11**(1), 76–77.

Martinson, B.C., Anderson, M.S., deVries, R. (2005). Scientists behaving badly. *Nature*, **435**, 737–738.

Nature Cell Biology (2006). Editorial: Appreciating data: warts, wrinkles and all. *Nat. Cell Biol.*, **8** (3), 203.

Nature Cell Biology. (2004). Editorial: Gel slicing and dicing: a recipe for disaster. *Nat. Cell Biol.*, **6**(4), 275.

Neroth, P. (2004). Tobacco ties. *Lancet*, **364**, 925–926.

Neville, C. (2007). *The complete guide to referencing and avoiding plagiarism*. Open University Press, Berkshire.

Odling-Smee, L., Giles, J., Fuyuno I., Cyranoski, D., Harris, M. (2007). Where are they now? *Nature*, **445**, 244–247.

Rinaldi, A. (2008). Access evolved? *EMBO Rep.* **9**, 317–321.

Rossner, M. and Yamada, K.M. (2004). What's in a picture? The temptation of image manipulation. *J. Cell Biol.*, **166**(1), 11–15.

Society of Authors (1965). *Quick quide: permissions* [online]. Available at: http://www.societyofauthors.net/ (last accessed 7 April 2008).

Suber, P. (2007). *Open access overview*. [online]. Available at: http://www.earlham.edu/~peters/fos/overview.htm (last accessed 9 May 2008).

UK Intellectual Property Office (2007). *Copyright basic facts* [online]. Available at: http://www.ipo.gov.uk (last accessed 10 October 2007).

University of Leeds (2005). *Plagiarism—University of Leeds Guide* [online]. Available at: http://www.ldu.leeds.ac.uk/plagiarism/ (last accessed 29 October 2007).

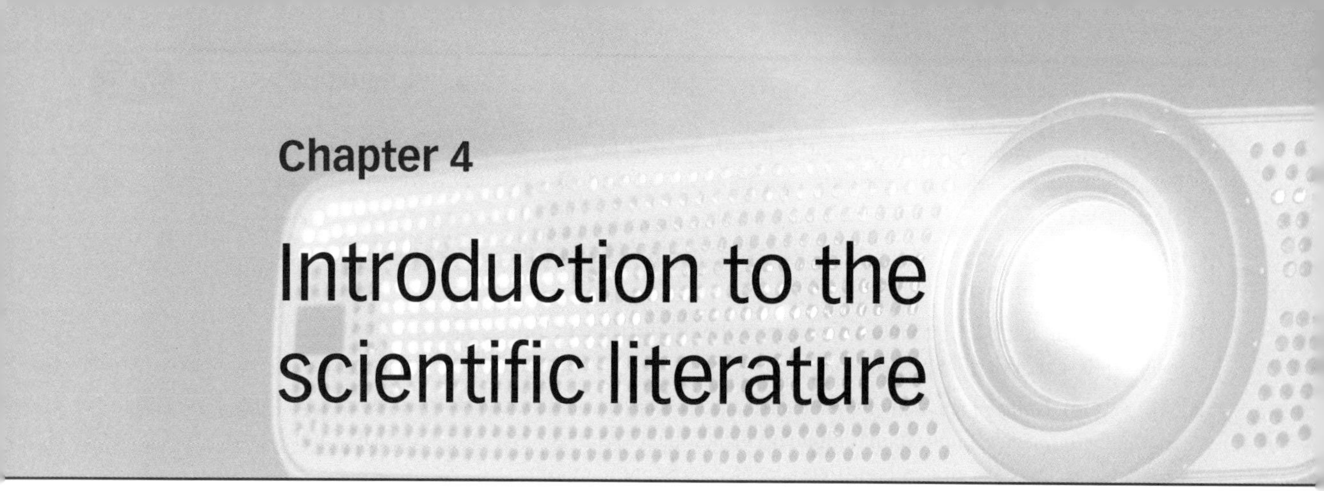

Chapter 4

Introduction to the scientific literature

➲ Introduction

Scientific material is communicated through a number of different channels including scientific journals, textbooks, theses, conference proceedings, and the worldwide web. These materials can be classified as primary or secondary sources of literature and as peer-reviewed or non-peer-reviewed literature. This chapter provides you with an overview of the different types of scientific publications and introduces you to the concepts of credibility and accountability in scientific publishing.

Specifically, this chapter will:

- define the terms primary and secondary sources, peer-reviewed and non-peer-reviewed work
- provide an overview of the different types of scientific publications
- outline the peer-review process
- discuss the journal hierarchy
- summarize the publication process for a journal article, from selecting a journal and consulting the instructions for authors, through to actual publication.

4.1 The main types of scientific literature

Scientific literature, as a body of material, is disseminated in a range of formats aimed at a range of audiences. However, the variety of publications that comprise the scientific literature all share common objectives. These are to be authoritative (that is, to be reliable sources of information) and to provide a contemporary account of the topic they cover. Beyond this, the scientific literature can be classified as primary or secondary literature and as peer-reviewed or non-peer-reviewed literature.

- The **primary literature** refers to those publications in which original ideas and data are first communicated. These are usually in the form of research papers (also called scientific papers) in peer-reviewed academic journals. Research papers form an essential part of scientific communication, as the findings reported in these papers represent new information which advances important conceptual or practical understanding of a particular problem.
- The **secondary literature** refers to those publications which report on, summarize, evaluate, or make some other use of information derived from the primary literature. An example of a secondary source is a review article (Table 4.1), which provides a critical overview of recent research in a particular topic area. Secondary sources of literature are typically aimed at a more general audience than the primary literature, and may digest the primary literature, presenting it in a format that is more readily accessible to those without particular expertise in the area being addressed.

A key objective of scientific publications is to provide information that is accurate, reliable, and current. In the scientific domain this is achieved through the **peer-review process** during which academic colleagues working in the same field as the publishing author evaluate the piece of writing before it is published. Literature that has been subjected to this evaluation process is referred to as **peer-reviewed literature.** However, not all scientific material that is available in the public domain is peer reviewed and it is likely that a large amount of material that you come across will be **non-peer-reviewed literature**—that is, material which has not undergone the same process of appraisal by academic peers as the peer reviewed literature has. The ability to differentiate between material that is peer reviewed and that which is non-peer reviewed is essential as it provides you with a starting point from which you can begin to assess the credibility of the information before making use of it in your own work. We will revisit the concept of peer review later on in this chapter when we describe the peer-review process (section 4.2) and again in Chapter 5 when we consider the various criteria we can apply to assess the academic integrity of various publications.

Scientific information can be disseminated in a range of formats. These include scientific journals, textbooks, monographs, conference proceedings, theses, official and technical publications (such as those produced by governments and their departments), patents, and web pages. The section that follows describes each of these publication types in more detail.

4.1.1 Scientific journals

Scientific journals are published periodically: either weekly, monthly, quarterly, or, in a few cases, annually. Scientific journals mainly comprise a series of articles reporting the most recent scientific developments. Journals may be broad in subject coverage, reporting developments across a wide spectrum of scientific disciplines—such as the journals *Nature* and *Science*—or may be highly specialized, focusing on a specific subject area—such as the journal *Neuroscience.*

A comprehensive list of journals published worldwide can be accessed from *Ulrich's Periodical Directory* (http://www.ulrichsweb.com/ulrichsweb), most of which are also listed on the Web of Science database (Chapter 5). Most of these journals are peer reviewed, but some are not. You can check whether a journal is peer reviewed or not, again by consulting *Ulrich's Periodicals Directory*. Peer-reviewed journals are considered to be more reputable than non-peer-reviewed journals as the papers in the former have been vetted for merit before publication. However, even amongst the body of peer-reviewed journals some are considered by scientists to be more prestigious than others. A number of factors are used to measure the prestige of a journal. These include numeric measurements such as the citation-based impact factor (see section 4.3), readership numbers, and rejection rates compared with other journals in the same field, as well as the reputation of the journal among the expert scientific community.

Scientific journals contain various types of articles whose exact definition may differ from journal to journal but broadly fall into the following categories: research (scientific) papers, letters, review articles, opinion pieces, editorials, letters to the editor, and conference proceedings. Some journals also include a news section as well as book reviews and reviews of multimedia publications. All of these article types are described in Table 4.1 together with an indication of whether they are primary or secondary sources of information and whether they are peer reviewed or not. You will note from the table that although most article types in academic journals are peer reviewed, a few are not; these include items such as editorials, letters to the editor, and news items.

TABLE 4.1 The different types of articles published in scientific journals

Article type	Description	Peer reviewed or non-peer reviewed?	Primary or secondary literature?
Research article (also known as the scientific paper)	Reports original research that has not previously been published elsewhere. These articles are descriptions of the nature of the problem studied, methodology utilized, comprehensive overview of the results achieved and the conclusions reached	Peer reviewed	Primary literature
Letter (or report)	Also reports original scientific research but shorter in length than the research article. Usually fast-tracked through the publication process as the findings are considered to be of sufficient importance to merit urgent publication	Peer reviewed	Primary literature

TABLE 4.1 Cont'd

Article type	Description	Peer reviewed or non-peer reviewed?	Primary or secondary literature?
Review article	Provides a critical overview of recent research in a particular field. Some journals are dedicated to publishing review articles such as the journal series *Trends in*… and *Current Opinion in* …. However, most if not all journals publish a proportion of reviews. Review articles can vary in length from short mini-reviews to longer, more extensive articles. Some review articles are commissioned by editors and some are submitted voluntarily	The majority are peer reviewed	Secondary literature
Opinion piece (include articles such as viewpoint, commentaries, and perspectives)	Brief articles which provide a fresh outlook or analysis of recent developments or topical issues that have recently been reported by other scientists	May not be peer reviewed by external reviewers but will be reviewed by the editorial team	Secondary literature
Editorial	Written by members of the editorial team. Topics discussed may include opinions on ethical and public concerns and recent developments in science	Non-peer reviewed	Secondary literature
Conference proceedings	Summaries of work presented at scientific conferences. These preliminary findings are usually submitted as complete research papers at a later date	Often peer reviewed if published in a journal	Primary literature
Book review	Describes and evaluates a newly published book. This section may also include multimedia reviews	Non-peer reviewed	Secondary literature
Letter to the editor/ correspondence	Response sent in by readers to previously published articles. Acts as a forum for discussion and debate amongst the readers and serves as a post-publication peer review mechanism which can improve the understanding of an article	Non-peer reviewed	Secondary literature
News section	Journals such as *Science* and *Nature* include articles presenting social and political issues that influence scientific policy	Non-peer reviewed	Primary or secondary literature (dependent on the news being reported)

Scientific journals are published by a range of organizations including societies, university presses, and commercial publishers. Some well-known publishers include Oxford University Press, Elsevier, Biomed Central, Blackwell/Wiley, and Macmillan. It is in the commercial interests of a publisher to make their journal as attractive as possible to a large readership; for example, by aiming for a high impact factor or by publishing a journal in a newly emerging discipline that is not covered by other publishers. It is also in the commercial interests of the publisher to publish as much scientific information as they can in the smallest space possible. As print space is limited, it is becoming more common for journals to include online supplementary material associated with the research papers they publish. This supplementary material can include information such as additional data or methodological details, which is not essential for understanding the study, but nonetheless the reader would benefit from reading.

4.1.2 Textbooks

Like scientific journals, textbooks can either be specialized, focusing on a specific topic, or cover a wide range of topics within a subject area. They can be written by a sole author or may be comprised of chapters written by multiple authors as invited by the editor. Textbooks are a good starting point for background reading as they provide an overview of basic facts and theories in a topic area. They may not, however, cover the most recent findings as they take a substantial amount of time to write and then to publish. The peer review of textbooks does not follow the exact procedure that journal articles are subjected to (see section 4.2). However, most (if not all) publishers will have draft chapters externally reviewed in a manner similar to the conventional peer-review process. Textbooks are generally secondary sources of information in that they review previous work on the subject, but they may occasionally include some primary literature.

4.1.3 Monographs

Monographs are highly specialized publications that are written on specific topics. They are distinct from textbooks in that they are not written to support teaching (although they can do so) but are written by specialists for specialists.

4.1.4 Conference proceedings

Conference proceedings are summaries of work presented at scientific conferences that are written up subsequent to the meeting. They are published either as a single bound conference volume, or as part of a journal publication, or as a supplement to the journal, and are peer reviewed. The findings presented at conferences are often from the early stages of research which has not yet been published as complete research papers. Such findings are therefore a valuable source of primary information as they provide a snapshot of the most current research taking place in a particular field. Most of these findings are submitted formally to journals as complete papers at a later date. Work presented at conferences is not always published as conference proceedings. Some conferences produce only an abstract book containing abstracts of work presented at the conference—these may or may not be peer reviewed.

4.1.5 Theses and dissertations

A thesis is usually defined as a major piece of original research submitted as a requirement for the award of a doctorate degree (e.g. PhD). A dissertation is usually defined as the original research submitted as a requirement for the award of a Master's degree. Both are primary sources of information that contain research findings, the most important of which are eventually published in journal articles. In addition, they contain a considerable review of the literature which can be useful. Theses and dissertations are not formally peer reviewed but have been peer reviewed in so far as the student must have satisfied the examiners in order for the thesis or dissertation to have been accepted.

4.1.6 Official and technical reports

This body of literature consists of reports (such as statistical reports, surveys, press releases, circulars, and legislation) that are published by government departments and other official organizations (such as the World Health Organization and the Royal Society). This category also includes technical reports such as manufacturer's protocols for kit and equipment use, produced by commercial suppliers of scientific products.

4.1.7 Patents

A patent is a legal document granted by the state to an inventor for a limited period of time. During this time period, no other individual is able to make, use, or sell the invention without the permission of the inventor. In return for this patent, the inventor pays a fee and discloses the technical details of their invention to the Patent Office, which are then published and made available online (see Chapter 5, Table 5.1). Patents are a highly specialized but very useful source of information.

4.1.8 Web-based resources

Over the years there has been a vast expansion in the volume of information that is readily available through the Internet. However, the quality of information is variable, ranging from high-quality academic sources to information that is factually incorrect, outdated, and misleading. You should, therefore, take great care when using web-based resources in your work and use only those that are peer reviewed or from authoritative sources. You can search for high-quality scientific resources using the following search tools:

- **Specialist academic databases,** such as Web of Science and PubMed, are searchable catalogues that store a range of scientific publications including journal articles and conference proceedings. These publications are organized (i.e. indexed) according to key terms that define the subject matter of that particular publication and hence allow researchers to retrieve useful publications quickly by typing in specific terms in the searchable boxes. The main biological sciences academic databases are listed in Table 5.2.

- **Data repositories** (or databanks) are sites where data associated with published research papers (or unpublished data) are deposited and stored. Examples of data repositories include the American Type Culture Collection (for depositing cell lines), GenBank (for DNA sequences), Protein Databank (for protein data), ArrayExpress (for microarray data), Knowledge Network for Biocomplexity (for ecology and biodiversity data), and the Global Populations Dynamics Database (for animal and plant population data).
- **Gateways** are an additional source of information. These maintain a catalogue of educational and research-based web resources that have been selected and vetted by academics for accuracy and quality. A list of biological sciences gateways is included in Chapter 5, section 5.2.3.
- **Specialized search engines** such as Google Scholar (a search engine for scholarly literature) and Scirus (a science-orientated search engine) exclude non-academic websites, and search for peer-reviewed papers, theses, books, abstracts, and other academic literature from a wide variety of (but not all) academic publishers, professional societies, and scholarly articles across the web. Literature extracted from these specialized search engines is therefore a more accurate source of information than literature obtained through general search engines such as Google and Yahoo! However, these search engines are significantly less comprehensive and much less useful than specialist academic databases (Shultz, 2007).

4.2 The peer-review process

4.2.1 What is peer review?

Peer review is a formal review process which is carried out before the material is published, during which qualified experts, who research and submit work for publication in the same field as the author, evaluate scientific work for significance, competence, and originality (Brown, 2004). The aim of peer review is to improve the quality of work by identifying mistakes and omissions at the preprint stage. Overall, the peer-review process is recognized by scientists as a valuable check mechanism for advancing a body of scientific knowledge that is both plausible and original. The review process can detect work that may be duplicated, flaws in experimental design, and poor analysis and interpretation of results. It cannot detect (and is not intended to detect) falsified or fabricated results (Lock, 1994) (see Chapter 3, section 3.3.3). However, in some instances the reviewers may suspect the authenticity of the results and alert the editor to this in their appraisal of the work.

4.2.2 How does peer review work?

This section describes the peer-review process which a research paper submitted to a journal for publication passes through.

Once a paper is received by the publishers, an editor assigned to the subject area evaluates the manuscript to see whether the study falls within the journal's scope and makes a decision on whether to send it out for peer review or not.

If, at this initial pre-screen stage, the editor decides that a submitted paper falls within the journal's subject area, s/he recruits two or more reviewers (also known as referees) who are experts in the same field (but independent of the authors and their institution) to review the work. Most papers operate the traditional closed peer-review system in which the identity of the authors and the reviewers is blinded from each other, or the identity of the author(s) is disclosed to the reviewer but not that of the reviewers to the author(s). This anonymity is intended to help reviewers make an objective and fair appraisal of the work without accusations of unethical behaviour. However, some journals are now moving towards a more open and transparent system of peer review where the identity of the authors and the reviewers is made known to each party. Some journals will also allow authors to suggest suitable reviewers and name those whom they would prefer not to be used. The arguments for and against open peer review are discussed in *Nature* (2006) as part of its peer-review debate.

Reviewers typically comment on the importance and originality of the work, appropriateness of the experimental design, whether the conclusions are supported by the evidence presented, and on the general structure, clarity, and length of the article. Typical review questions are:

1. What is novel and significant about this work?
2. Is the methodology and statistical analysis sound?
3. Are the claims justified by the results or is additional work required to support the claims the paper is making?
4. If additional work is required, what are the experiments?
5. Are the claims appropriately discussed in the context of the published literature or are there particular references that are missing and should be included?
6. Is the paper clearly and comprehensively written or are there parts that should be deleted or expanded?
7. Are there any other comments you would like to make?

Based on their appraisal of the work, the reviewers may make one of the following five recommendations to the editorial team:

- Accept the paper for publication unconditionally without any modifications.
- Accept the paper for publication but only after minor modifications have been made.
- Reject the paper in the present form but suggest that with major modifications to the existing work, it may be suitable for resubmission to the same journal.

- Reject the paper but encourage submission to another journal.
- Reject the paper unconditionally, as the work has serious defects.

Based on the responses received from the referees, the editorial team will make a judgement on whether to publish the work or not. If the reviews are contradictory, the editorial board may seek advice from additional referees.

If the outcome is to publish with modifications, the editor will write to the authors with the feedback and recommendations received from the reviewers. The authors may accept all of the suggestions made for improvement and revise the paper accordingly or, if they do not agree with aspects of the referees comments, counter-respond with reasoned arguments stating why the suggestions have not been implemented. The revised manuscript (and/or reasons for rejecting the suggestions) may be sent out for reassessment before going to print.

For peer review to work well, authors must present their work honestly and concisely, and this is discussed in Chapter 3 (section 3.3.3). Similarly, reviewers must provide honest, constructive, and timely feedback on the papers they review. Therefore, the key characteristics of a successful peer review are:

- **Expert review:** reviewers are selected based on their detailed knowledge of the subject area. This is essential, as the reviewer is expected to pick up duplicated work, poorly designed experiments, and poor analysis and interpretation of results.
- **Confidentiality:** reviewers are expected to maintain confidentiality about the content of the paper they are reviewing and not discuss the work with other people unless permission from the editors is granted; for example, when a colleague is assisting in the review process.
- **Objectivity and fairness:** reviewers are expected to be objective and fair in their appraisal of the work. This means not allowing any sources of bias (personal, financial, or professional) to influence the evaluation. An example of when bias may creep in is if the work under review relates directly to that of the reviewer. In this case the reviewer may be tempted to block the paper by providing a discouraging review of the work so that his/her research group can publish the work first. The scientific community expects that the reviewer will not undermine the integrity of the peer-review process by acting dishonestly in such cases.
- **Timely return of the feedback:** reviewers are typically given 2 weeks to review and submit their report on a manuscript. Timely return of feedback is important so that the work can be published and made available to the public.

Once published, peer-reviewed work becomes part of the scientific record and is recognized as work that has scientific value and integrity. However, the peer-review process does not stop at the point of publication but the published work continues to be scrutinized and assessed by the scientific community. This is through a variety of channels such as discussions at journal clubs, letters to the editor, and debates at conferences. This post-publication discussion is central to the progress of scientific knowledge as it allows independent scientists to evaluate and confirm tentative discoveries or, in certain instances, discount them (Chapter 6).

4.3 The journal hierarchy

As noted in section 4.1.1, the quality and prestige of a journal is evaluated using either subjective methods such as the opinions of experts in the academic field, or objective methods such as acceptance rates and journal impact factors. Of these methods, the journal impact factor is now commonly used as a numeric measurement for ranking journals hierarchically. It is calculated annually by the Institute for Scientific Information (ISI) and published on the Journal Citation Reports (JCR) database.

The impact factor for a particular journal in a specific year is defined as the number of times articles published in this particular journal are cited in the previous 2 years divided by the total number of articles published in this journal in the previous 2 years (excluding article types such as news items, correspondence, and editorials) (Garfield, 1996; 1997).

To illustrate, the impact factor for the journal *Cell* for 2007 would be calculated as follows (ISI, 2007):

Citations in 2007 to articles published in 2006 (in journals tracked by ISI) = 10096

Citations in 2007 to articles published in 2005 (in journals tracked by ISI) = 9958

Total citations = 20054

Number of articles published in *Cell* in 2006 = 352

Number of articles published in *Cell* in 2005 = 319

Total articles = 671

Impact factor for 2007 = total citations/total articles = 20054/671 = 29.887

The values of impact factors can range from zero for the least cited to values greater than 50 for some of the most highly cited journals. Table 4.2 lists the top 10 scientific journals ranked by impact factor in 2007 (see page 60).

4.3.1 Note of caution

Impact factors as a tool for measuring the quality of a journal should be used with caution as the number of citations of a journal can be inflated by a number of factors such as the popularity and size of the research field (that is, the size of circulation of the journal) and the number of review articles included in the journals (as these are cited frequently) (Jones, 2003; Wilson, 2007). Therefore, journals with a wide range of topics and hence a wider readership will have higher impact factors than specialized ones. A low impact factor does not necessarily mean that the quality of the journal is poor. For example, the *Journal of Dermatology* has a low impact factor but is the best journal in the field of experimental dermatology. In this instance, the low impact factor is a reflection of the more specialized nature of the journal and hence its smaller readership. It is therefore important that you consider the position a particular journal holds within a field when evaluating the quality of a journal. It is also worth noting that the

TABLE 4.2 The top 10 scientific journals ranked by impact factor in 2007 (ISI)

Rank	Journal title	Impact factor
1	*Cancer Journal for Clinicians*	69.026
2	*New England Journal of Medicine*	52.589
3	*Annual Review of Immunology*	47.981
4	*Reviews of Modern Physics*	38.403
5	*Nature Reviews Molecular Cell Biology*	31.921
6	*Annual Review of Biochemistry*	31.190
7	*Cell*	29.887
8	*Physiological Reviews*	29.600
9	*Nature Reviews Cancer*	29.190
10	*Nature*	28.751

impact factor is a reflection of the citation rate of an *average* article in a journal, not of a specific article, and therefore cannot be used to measure directly the scientific worth of an individual paper.

4.4 The publication process

This section describes the process of publishing scientific work in peer-reviewed journals starting with selecting a publication outlet for your work and then consulting the instructions for authors, through to actual publication.

4.4.1 Selecting a publication outlet for your work

Before starting to write a paper for publication, whether a research article, review article, or any other, you should have some idea of which journal you would like to publish in. Factors which may influence your choice of journal are (adapted from Borgman, 1993):

- **Reputation** of the journal (see section 4.3).
- **Scope and readership:** some journals publish only fundamental breakthroughs in research. Some publish a wide spectrum of articles, whereas others accept papers from within a highly specialized field only. You will need to consider whether the journal scope and readership match the content and significance of your work and the audience you want to target.

- **Acceptance rate:** some journals are highly selective and therefore reject a large number of the articles they receive (for example, *The Lancet* rejects over 90% of articles received). This does not mean that the rejected work is not of a good quality. Most papers that are rejected go on to be published in other journals which have a different scope or are of lower ranking. You will therefore need to consider the significance and scope of your work and whether this aligns with the journal choice. If your work does not match the significance and scope of the journal then the likelihood of your paper being rejected will be high.
- **Turnaround time:** the time taken from receipt of the paper to its appearance in print varies from journal to journal and can take from 6 months to over a year. Clearly, a faster turnaround time will ensure that your paper will be in the public domain much faster.
- **Open access or subscription-only journal:** for open access journals, publishing charges are borne by the author(s) (that is, their institution or research sponsors). This allows the public to access the article free of charge. In subscription-based journals; the journal pays for the publishing costs (but see the following point) and the article is restricted to subscribers only. Some journals now give authors the option to publish their article as open access or through the regular subscription model within the same journal. As an author your decision to publish your work via the open access model or the subscription model may be determined by a number of factors (such as personal motivations and the requirements of funding organizations) which were discussed in Chapter 3 (section 3.3.2).
- **The cost of publishing colour images and page charges:** colour illustrations are expensive to produce and if your paper includes such illustrations some journals may ask for a fee to reproduce them. Journals can also ask for page charges (for each page of the manuscript) or excess page charges (for every page that exceeds the length of manuscript stipulated by the journal). The fees charged may affect your decision to publish in a particular journal.

4.4.2 **Instructions for authors**

Once you have decided where to publish, you need to familiarize yourself with the instructions for authors supplied by the publishers, either online or in an issue of the journal. Although similar formats and conventions apply to specific article types (for example, all research articles will include Abstract, Methods, Results, and Discussion sections), precise details vary from journal to journal. These details apply to overall length and layout, title page, structure and sections of the article, figures and tables (number, type, and quality), referencing style, and format of submission. It is essential that as a prospective author you are aware of these rules and conform exactly to them. If not, your article will be either rejected or returned for rewriting in the correct journal style. An example of instructions for authors supplied by Oxford University Press for the journals it publishes can be viewed online at the journal homepage (http://www.oxfordjournals.org/for_authors/).

4.4.3 From submission to publication

Once the article is submitted, it is either rejected or sent out for peer review (see section 4.2). If it is then accepted for publication, the final accepted manuscript is copy edited in line with the style of the journal and assigned to an issue for publication. The entire process from receiving a first draft of a manuscript to final publication in print usually takes about 6 months. The time scale may be shorter or longer depending on the speed with which editorial decisions are made, reviewer's reports are received, and amendments to the draft manuscript are returned by the authors. If findings are considered high priority, a paper may be fast-tracked through the publication process.

Most journals now make the revised accepted manuscript available online to allow readers access to the paper before it is published in a journal issue. At this stage the article is assigned a unique Digital Object Identifier (doi) number (http://www.doi.org/) so that the paper can be referred by it before it is formally published. Publishers use these numbers to identify and to provide a persistent link to individual electronic publications which may not always have volume and page number information.

References

Borgman, C. L. (1993). Round in circles: the scholar as author and end user in the electronic environment. In Woodward, H. and Pilling, S. ed. *The Internationals Serials Industry.* Gower Publishing, Aldershot, Hampshire.

Brown, T. (2004). *Sense about science. Peer review and the acceptance of new scientific ideas* [online]. Available at: http://www.senseaboutscience.org.uk/index.php/site/project/29/ (last accessed 9 November 2007).

Garfield, E. (1997). Dispelling a few common myths about journal citation impact factors. *Scientist*, **11**, 11.

Garfield, E. (1996). How can impact factors be improved? *Br. Med. J.*, **313**, 411–413.

ISI (2007). Institute for Scientific Information. *Journal Citation Reports* [online]. Available at: http://wok.mimas.ac.uk/ (last accessed 22 June 2008).

Jones, A.W. (2003). Impact factors of forensic science and toxicology journals: what do the numbers really mean? *Forensic Sci. Int.*, **133**(1–2), 1–8.

Lock, S. (1994). Does editorial peer review work? *Ann. Intern. Med.*, **121**, 60–61.

Nature (2006). *Nature*'s peer review debate [online]. Available at: http://blogs.nature.com/peer-to-peer/categories/peer_review_debate/ (last accessed 28 May 2008).

Shultz, M. (2007). Comparing test searches in PubMed and Google Scholar. *J. Med. Libr. Assoc.*, **95 (4),** 442–445.

Wilson, A.E. (2007). Journal impact factors are inflated. *BioScience*, **57**(7), 550–551.

Chapter 5

Conducting effective literature searches

➲ Introduction

The aim of this chapter is to equip you with the necessary skills to construct and implement a systematic search of the literature and retrieve information which is most relevant to the topic of your search in a time-efficient way.

Specifically, this chapter will:

- describe the key features of an effective literature search
- describe the search tools you can use to locate particular types of publications and how to use these tools effectively
- outline a strategy for planning and implementing a comprehensive literature search
- describe how you can keep track of scientific literature using alerting services and RSS feeds.

5.1 The literature search

The ability to search and retrieve particular types of information is an essential part of academic work. You will therefore conduct literature searches often and for a variety of reasons. For example, if you are new to a particular topic, you may conduct a search to obtain a general overview of the area. If you are already familiar with the topic, then you may conduct a literature search to locate the new developments in the field. Alternatively, you could be attempting to find information about a specific method or protein or organism. The reasons for your search will determine which publication types you retrieve and which search tools you use. Therefore, before embarking on any literature search, ask yourself the following three questions:

1. Why am I searching the literature?
2. What types of information am I looking for?
3. Which search tools should I use to locate the information?

Your responses to these questions will help you plan your literature search and are discussed further in section 5.4.

When searching the literature, you must be effective. That is, you must be able to scan the large amounts of information and locate the sources which are most relevant to the topic of your search in a **time-efficient way**. To achieve this:

- Be **systematic** in your approach. This means you must proceed through a series of logical steps, which include first planning and then implementing your search, to retrieve the relevant information.
- Be **comprehensive** in your search. This means you must search widely so that you do not omit any important information.
- **Select** your sources of information carefully. This means you must use high-quality sources which are factually correct, up to date, and reliable. Your major source of information should be peer-reviewed scientific articles published in journals. However, authoritative information such as patent details, government statistics, research legislation, and manufacturer's protocols are also appropriate sources of information.

Chapter 4 describes the different formats in which scientific information is disseminated and the section that follows describes the search tools you can use to locate particular types of publications and how to use the tools effectively.

5.2 Tools for searching the literature

Table 5.1 lists the different types of scientific publications and the corresponding search tools that can be used to locate the information. You will see from the table that the main search tools are library catalogues, specialist databases, and gateways.

5.2.1 The library catalogue

Depending on the extent of holdings, the library catalogue can be used to search for a whole range of material types including textbooks, journals, theses, patents, conference proceedings, encyclopaedias, and audiovisual material. The catalogues generally use the principle of keyword, author names, title names, or ISBN number searches to help retrieve the materials. These publications may be available as hard copies, electronic copies, or a combination of both.

Before embarking on any literature search, familiarize yourself with the types of literature your library holds and how the search facilities are organized. Most university libraries have subject-specific librarians who will be able to advise you. Libraries may also provide face-to-face training sessions on information seeking and retrieval skills or have online tutorials that you can work through independently. In addition to your university library it may be possible to extend your search to other libraries, both national and international (Table 5.1).

TABLE 5.1 Types of publications and the corresponding search tools that can be used to locate them

Publication type	Search tool
Journal articles	Specialized electronic databases (Table 5.2)
Textbooks and monographs	**Library online public access catalogues (OPACs)** A list of library OPACs can be accessed via the gateway BUBL (5.2.3) and includes: *National library OPACs* such as: Copac Academic and National Library Catalogue—this catalogue gives free access to the online catalogues of the major university and national libraries in the UK and Ireland The British Library Public Catalogue *International library OPACS* such as: LibDex—this catalogue lists 17 000 worldwide libraries **National bibliographic databases** (such as the British Bibliographic Database and the Australian Bibliographic Database) and web-based services that search publishers' catalogues (e.g. Amazon) can also be used to search for textbooks and monographs
Conference proceedings	Conference papers index database (part of the CSA database) (Table 5.2) ISI proceedings database (via Web of Knowledge) (Table 5.2)
Theses and dissertations	Theses are held by the university at which the work is undertaken. National and international theses and dissertations can be located by searching through the following databases: Index to Theses—includes abstracts from British and Irish Universities ProQuest Dissertations and Theses—includes details of worldwide PhD theses and Master's dissertations Networked Digital Library of Theses and Dissertations (NDLTD)—provides access to full-text electronic theses mainly to those awarded by USA institutions (but participation by other countries is growing)
Technical and official reports	Available from corresponding corporate websites
Patents	European Patent Office database and USPTO (US Patent and Trademark Office) database—both contain bibliographic and full-text details of patents

5.2.2 Specialist electronic databases

Electronic databases are searchable catalogues that store a range of publications including journal articles and conference proceedings. These publications are organized (i.e. indexed) according to key terms that define the subject matter of that particular publication and hence allow researchers to retrieve useful publications as quickly as possible by typing in specific terms in the searchable boxes.

There are two main types of electronic databases: bibliographic databases and full-text databases. The former contain citation details (author(s), article and journal title, date of publication, volume and page numbers) and usually abstracts for all the articles indexed; the latter contain the full text for the articles indexed. Complete journal articles listed in most databases are available only on subscription. However, you will be able to access the full text if it is published under open access or free access (Chapter 3, section 3.3.2). An increasing number of journals are now making the full text of articles available to the reader after a fixed time, often 1 year after publication, and you will be able to access the these articles as well (section 5.4.2).

Table 5.2 lists the most commonly used specialized electronic databases in the biosciences.

TABLE 5.2 The most commonly used databases in the biosciences

Database	Description
Biosis Previews	Indexes more than 6000 journals and covers all areas of biological research. Allows search of research articles as well as reviews, monographs, conferences and meetings, and technical reports. Coverage from 1969 onwards
CAB Abstracts	Allows access to abstracts from more than 9000 journals in the areas of agriculture, forestry, human nutrition, animal health, and the management and conservation of natural resources. Coverage from 1910 onwards
Cambridge Scientific Abstracts	A collection of databases that allows access to abstracts and citations in the areas of biomedicine, biotechnology, zoology, and ecology, and some aspects of agriculture and veterinary science. Allows search of 6000 journals as well as conference proceedings, technical reports, monographs, books, and patents. Coverage varies depending on the database but can be from 1960 onwards
Medline	Abstracts of articles from more than 4900 journals in areas of medicine, dentistry, and nursing. Coverage from 1951 onwards
PubMed	Allows search of over 5000 biomedical journals including those in the Medline database as well as biomedical articles from further life science journals. Coverage from 1950 onwards
Science Direct	Access to the full text of over 1800 journals published by Elsevier. Subjects include science, medicine, and engineering, as well as business, management, and social sciences. Coverage from 1995 onwards
Web of Science—Science Citation Index	Science Citation Index is available from within the Web of Science database which is part of the ISI Web of Knowledge database. Allows search of more than 8700 peer-reviewed journals in the fields of natural and biomedical science and technology. Allows access to abstracts and references and some links to full-text articles. Coverage from 1900 onwards
Zoological Records	Covers publications in all areas of zoology and animal sciences, and includes references from over 4500 journals as well as books and meetings. Coverage is from 1978 onwards

5.2.3 Gateways and other web-based resources

Subject-specific gateways, also known as portals, contain links to web resources that have been selected and evaluated by acknowledged experts in the field. Some gateways are highly specialized and contain links to resources in a particular subject area only. Other gateways are less specialized and contain links to a wide range of educational and research-based web resources.

Examples of highly specialized gateways are:

- **Cell Signalling Gateway** (http://www.signaling-gateway.org/)
- **Cell Migration Gateway** (http://www.cellmigration.org/index.shtml)
- **Neuroscience Gateway** (http://www.neuroscience-gateway.org)
- **Omics Gateway** (http://www.nature.com/omics/index.html)
- **National Biodiversity Network** (http://www.nbn.org.uk/).

These gateways attempt to provide an integrated view of the subject area by summarizing information from different sources and providing links to research, news, and events in their respective fields. For example, the cell signalling gateway contains links to important peer-reviewed research publications in the field and to bioinformatics databases. It also includes molecule pages which contain key information on over 3500 proteins involved in cell signalling—summarized from external databases—and includes DNA and protein sequence information, and structural and biochemical properties. Other specialist gateways similarly summarize and provide links to information in their respective subject areas and are an excellent resource.

Examples of less specialized gateways that contain links to a wide range of educational and research-based material include:

- **Intute** (http://www.intute.ac.uk/). This free online database contains information divided into subject-specific areas and includes research publications and news items, as well as information on software, organizations, and legislation.
- **BUBL** (http://bubl.ac.uk/). A catalogue of free online resources again divided into different specialist subject areas. It includes links to research publications, learned societies, genome maps, and databases.
- **World Wide Web Virtual Library** (http://vlib.org/). A free online library that again contains a wide range of subject-specific information including research publications; directories of research institutes and pharmaceutical companies; genomic, drug, and bioinformatics databases; software information; data standards; and technical and product information.

Evaluating web-based resources

A large amount of information you encounter on the web will be factually incorrect, outdated, and misleading, and therefore will not be suitable for academic use. A specific example of a **source that is not acceptable for academic work** is Wikipedia. Wikipedia is a searchable, online free-content encyclopaedia. It contains secondary information on a vast range of topic areas, and is authored by its readers voluntarily from around the world. A formal peer-review mechanism for checking the quality and accuracy of the material does not exist and very little is known about the authors—who may or may not be experts. This makes Wikipedia an unacceptable source of information for academic work.

The ability to differentiate between authoritative web-based sources and those which are not trustworthy sources of information is an essential skill which you should develop. You can assess the quality of a source by applying a set of stringent critical questions to the material you read. These questions are (see also Open University, 2008):

- **Is the material accurate?** The material should be fully referenced so that the information contained within the article can be verified through cross-checking with the original sources on which the information is based. From these original sources you will be able to determine whether the material you are reading is based on peer-reviewed work or not and hence how accurate it is likely to be.
- **How up to date is the information?** You will need to check when the information was last updated. Many web pages are out of date and therefore it is possible that the information you are reading has been superseded since it was last updated. If it has been superseded, then depending on the type of information and its relevance to your topic area, it may no longer be suitable for use.
- **Who is the author of the information and what are his/her qualifications?** Find out who authored the pages and whether they are qualified in the area they are writing about. You can find out this information by checking to see whether they work for a reputable institution or not. You can also verify the reputation of the author by checking whether they have published anything else and whether other sources have cited their work. This can be done relatively easily by conducting an author and citation search on a database such as Web of Science. If the web page is authored by an organization, then try to find out how credible the organization is (see below). Do not use the information unless you can verify the credentials of the author and/or the organization.
- **Who is the publisher of the information?** The URL of a web page can provide some information about the publisher. For example, URLs ending in **.com** are commercial sites; those ending in **.org** are non-profit organizations, those ending in **.gov** are US government sites, and **.gov.uk** are UK government sites. Academic sites are suffixed by **.ac.uk** (for UK universities and colleges), **.edu** (for academic institutions in the USA), and **edu.au** (for academic institutions in Australia). Publications by academic institutions, governments, and non-profit organizations are likely to be reliable sources of information. Commercial suppliers of scientific research products also publish product and protocol information which is of a high quality and hence suitable for use. Regardless of who the publisher is, all information must be assessed for objectivity (see below) before use.

- **Is the information objective?** The site should present the information in a balanced and objective manner with any assertions made supported by factual evidence which can be cross-checked against the original source. Some publications may express a particular perspective, for example in controversial areas such as stem cell technology or genetically modified crops. This type of open bias is easy to identify. What you need to be aware of is hidden bias which could intentionally mislead the reader. One way of identifying potential bias is to identify the purpose of the website and then determine whether the publisher may have a vested interest. For example, a drug company maintaining a website about their products may have financial vested interests; a government site may attempt to influence public opinion; or a conservation charity using its web pages to promote its projects, report research, and raise money may attempt to influence public opinion in order to raise funds. Make sure you consider any sources of bias prior to using the information.

5.3 Search functions of electronic databases

Bibliographic and full-text electronic databases have a number of search functions to help you locate the most relevant information. These functions include keyword searches; refining searches by setting limits; combining search terms; searching by author, journal title, or article title; and citation searches. Different electronic databases (Table 5.2) may use different conventions, but all have similar search facilities. For specific details, you will need to consult the help pages on each individual database which will guide you through the use of the database (for example, see ISI, 2008).

5.3.1 Keyword searches

Electronic databases allow you to retrieve relevant publications by typing keywords in a search box. When you do this, a list of articles which match your search term will appear on the screen with citation details for each article. If you click on the title it will link you to the abstract and in some cases the full text of the article.

When using search terms, consider using **truncations** (also known as **wild cards**). This feature allows you to search a word with alternative endings (that is, the same root). For example, searching for **enzym*** will retrieve items containing **enzyme, enzymes, enzymology,** or **enzymatic.** Variations of words with alternative spelling in the middle can also be retrieved by using truncation embedded within the word. For example, by searching for **fertli*e**, you will retrieve items with the variant spellings **fertilize** and **fertilise**.

5.3.2 Setting limits

You can refine your search by setting limits. Limits can be set to search certain languages, certain article types, and certain years of publication, as well as different types of studies.

5.3.3 Combining search terms

To make your search more specific you can combine search terms using Boolean operators; that is, words such as **AND, OR**, and **NOT**. For example, you may initially search for the terms **MRSA** and **antibiotics** as single keywords. You may then attempt to narrow your search by combining the terms **MRSA AND antibiotics**. This will reduce the number of results but the items you locate are more likely to be relevant to your topic. This is shown in the form of a Venn diagram in Figure 5.1. If you consider the MRSA circle as containing a set of references and the antibiotics circle as a set of references, then the overlapping area (highlighted in blue) are those references that contain both the terms MRSA and antibiotics and it will be these that are retrieved.

A way of expanding the search is to use the Boolean operator **OR**. This is useful if you are using search words with, for example, alternative spellings. To illustrate, you may want to search for MRSA but also for methicillin-resistant *Staphylococcus aureus*. Searching for **MRSA OR methicillin-resistant *Staphylococcus aureus*** will retrieve all documents that contain either of these two terms. Again this is shown in the form of a Venn diagram (Figure 5.2). The documents that fall in the highlighted areas, blue and grey, will be retrieved. This type of search will return a higher number of results but these may be of lower relevance.

The Boolean operator **NOT** allows you to exclude words from searches to prevent resources that include a particular word from being retrieved. For example, you may want to conduct the search **MRSA NOT VRSA**, which will retrieve items that contain the term MRSA but exclude those that contain the term VRSA (vancomycin-resistant

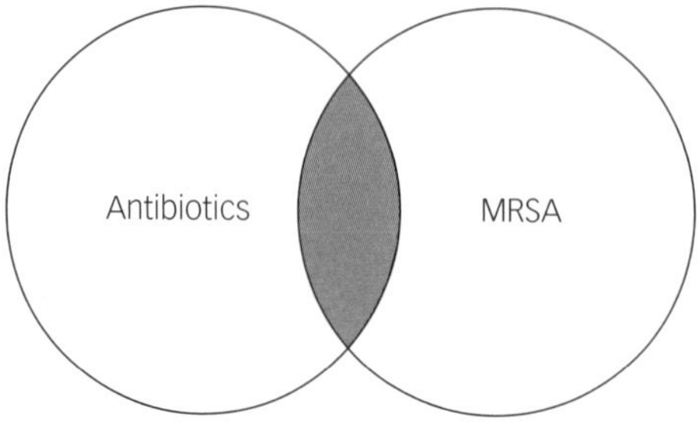

FIGURE 5.1 Use of the Boolean operator **AND**: results of searching for MRSA **AND** antibiotics.

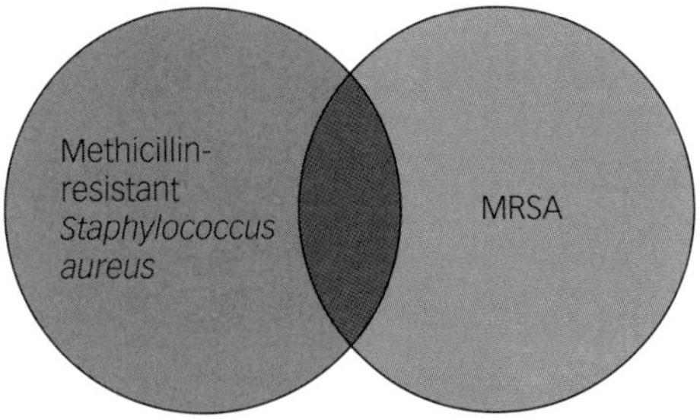

FIGURE 5.2 Use of the Boolean operator **OR**: results of searching for MRSA **OR** methicillin-resistant *Staphylococcus aureus*.

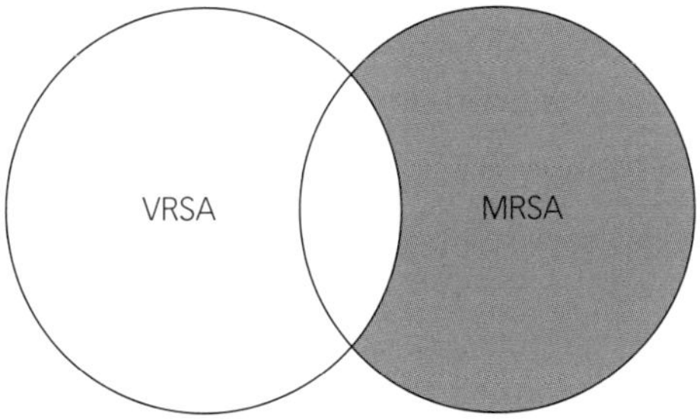

FIGURE 5.3 Use of the Boolean operator **NOT**: results of searching for MRSA **NOT** VRSA.

Staphylococcus aureus). Again, the documents retrieved are highlighted in a Venn diagram (Figure 5.3). The results will be fewer but of a higher relevance.

5.3.4 **Advanced searching**

By using the advanced search tool you can put more complex searches together by using more than one Boolean operator and thereby allow many smaller searches to be linked together. This can become quite complicated but if done properly can make your search highly specific and save you much time. Advanced searching also allows you to search for specific resources by author's names, article titles, title words, journal titles, or a combination of these.

5.3.5 **Citation searches**

If you have located a useful paper and would like to retrieve a list of other articles which have cited that paper in their work, then you can undertake a citation search. Citation searches are an extremely useful way of locating new papers that cite an older paper, thereby widening your literature search in a specific topic area.

5.4 An effective literature search

The way you go about conducting your literature search will depend on the purpose of your search. Therefore, you must have a clear idea of:

- why you are searching the literature
- what types of information you are looking for
- where this information is located.

To illustrate this:

- If you are writing a literature review on a new topic, you will be attempting to acquaint yourself with the background to the topic area. Your principal source of information will therefore be primary and secondary journal articles and you will search the electronic databases to locate the articles.
- If you are already familiar with the topic area, you may be conducting a search to find out about any new and recent developments in the field. In this case, you may restrict your search to conference proceedings, patents, and the primary articles published in the past few weeks or months.
- If you are searching for information on a particular protein, your sources of information may be DNA and protein databases and specialist gateways.
- If you are searching for information on legislation that governs research, then your sources of information may be government websites such as the Department of Health.

This section demonstrates how you can plan and implement a comprehensive and systematic literature search if you are **attempting to familiarize yourself with a topic that is new to you**.

5.4.1 Planning and implementing your literature search

If you are new to a topic area, then your first task is to identify one or two starter references which will give you a general overview of the basic facts and theories in the topic area. Review articles are very useful for this. You could ask your supervisor for suitable starter references or you could conduct a search within a database such as Web of Science or PubMed or another suitable database (Table 5.2).

- Before you start searching the database, formulate the search terms you will use. List key words associated with the topic you are researching. Consider alternative spellings and endings and consider combining the search terms to narrow or widen your search (see section 5.3). Restrict this initial search to review-type articles and limit it to current publications, for example review articles published within the past 5 years.
- Once you have read the initial references you are ready to widen the scope of your literature search and there are a number of ways you can do this. One is to use the

'ancestry approach'. This approach retrieves information by tracking citations from one paper to another. For example, you can use the reference lists in your starter review articles to track down further suitable references and from these papers track down further references, and so on. This works backwards, so that you retrieve older references. However, you can retrieve newer articles that cite the older reference by using the citation search facility in the databases (see section 5.3). Some databases such as PubMed have a 'related articles' facility, which links you through to articles related to your reference, and this is also a very useful way of finding articles related to a particular topic area.

- When searching databases do not confine your search to a single database; instead, carry out searches of multiple databases. This way you avoid missing key studies relevant to your topic.
- Follow up broad preliminary searches with further searches which are narrower and much more defined. You will find that during your initial searches you are exploring the full extent of literature that exists in the field and hence these searches will be relatively broad in scope. As you conduct further searches, these should become narrower and much more specific, addressing particular components of the topic. At this stage you should be formulating more complex search terms and defining your search parameters to narrow the search. Conducting author searches is very useful at this stage if you have identified a key author who is working and publishing in the field.
- Each literature search will yield a list of citations, but not all of these may be relevant to your topic. You will therefore need to make a judgement about which papers are related to your topic area and therefore useful for you to read in full. Reading the title and abstract of each citation will help you judge the relevance of the article content in relation to your topic area. A high number of citations for a particular article is a good indicator of an interesting article, and such references are also worth looking into.
- Once you have identified the publications you want to read you will then need to get hold of the complete document. If the article is open access then you can obtain the article directly through the database by clicking on the full-text icon. However, you will find that most articles listed in the databases are available only on subscription. If your library has paid a subscription fee for these journals then you will be able to access the complete article electronically, either directly through the database or through your library. If not, you may still be able to access the document through an inter-library loan.

5.4.2 Managing your search history and your references

It is essential to keep an accurate and complete record of your search history, as you may need to go back and undertake further searches and/or you may need to write up the details of your search strategy as part of your work. Some databases (such as Web of Science) will allow you to save your search histories within the database server itself. Alternatively, you could record the details of your search history by hand into a notebook or make a printout of your search screens and results. Whichever system you use, you must record the details of the search history in full, including the database searched, the limits set, the search terms used, and the date you conducted the search.

Once you have identified a list of likely references you will use, keep a record of them. This can be done easily using a reference management software package such as EndNote which enables you to download references directly from the electronic databases into a bibliography. You may also want to consider storing your references online on sites such as Connotea or CiteUlike (see Chapter 2, section 2.3). Alternative methods for keeping records of your retrieved references are to print the citation results from the screen directly, or, depending on the facilities available, e-mail the results to yourself or save them to a file on your personal hard drive. It is best to explore the facilities of the individual database to see which options are available to you and then use the one that is the most efficient—downloading references into reference management software or storing them on online sites are usually the most efficient ways of managing your references.

5.5 Tools for keeping up to date with the literature

Once you have set up a search or identified a paper, journal, or website that is of particular interest to you, then you can keep track of new information as it is published using tools such as alerting services and RSS feeds. These two tools are described below.

5.5.1 Alerting services

Alerting services work by sending you an e-mail when new information matching your alert criteria becomes available. To begin with you need to set up an alert by registering on the site of interest. The process of registering is quick and easy and involves setting up the alert criteria, entering your e-mail address, and registering a password. There are different types of alerts you can set up:

- **Search term alert**. This type of alert allows you to request an e-mail each time an article is published that matches your saved search terms. The search terms could be keywords, selected journal titles, or an author's name. Search term alerts can be set up in databases such as Web of Science, Cambridge Scientific Abstracts, and Medline. Some journal publishers such as Oxford University Press, Nature Publishing Group, and Elsevier also provide a search term alert service for the journals they publish.

This is an extremely useful service that can save a substantial amount of time as you do not have to repeat your searches, and is worth setting up for topic areas that you need to keep up to date with.

- **Journal article citation alert**. This type of alert allows you to request an e-mail each time your selected journal article is cited by new articles. Citation alerts can be set up within databases such as Web of Science, and some journal publishers such as Oxford University Press, Nature Publishing Group, and Elsevier also provide a citation alert service. The advantage of creating a citation alert within a database such as Web of Science is that you will be alerted when any of the journals listed within the database cites the article. If you set up an alert with a journal publisher, then the alert will be limited to citations by the journals published by that particular publisher. The article citation alert is very useful if you are interested in tracking who is citing a particular article. This may be a key paper in your subject area or a paper you have published. In both cases, you will be able to keep up to date with new publications related to your article of interest and find out which researchers are working in the same field as you.
- **Table of contents alert for a journal**. This type of alert allows you to request an e-mail each time a new issue of a journal is published. The e-mail will list the contents of the journal and may also include links to the abstract and the full text of the articles. Table of content alerts are provided by journal publishers such as Oxford University Press, Elsevier, Nature Publishing Group, BioOne, and Cambridge University Press, and can be set up through the journal title homepage. If you read a particular journal regularly, then this is worth setting up. Table of contents alerts can also be set up through a Zetoc alert (http://zetoc.mmas.ac.uk/). Zetoc provides access to the British Library's electronic table of contents for 20 000 journals and around 16 000 conference proceedings. It has the advantage of presenting a wider range of titles from different publishers and hence you can set up alerts for multiple journals from more than one publisher using a single alerting service.

5.5.2 RSS feeds

The vast majority of websites, including academic publishers and databases such as PubMed, include RSS (Rich Site Summary or Really Simple Syndication) feeds on their sites. This feed supplies you with the latest updates made to the site. The information is sent to a feed reader from where you read the material. The advantage of RSS feeds is that you can track the content of a large number of websites without having to browse the sites regularly, and all the updates from your selected websites are stored and read from a single location. There are two simple steps to receiving an RSS feed. The first is to register with a feed reader. A list of useful readers can be accessed from the Open Directory Project (http://www.dmoz.org/). Once the feed reader is installed, you can add an RSS feed into it by visiting your site of interest, clicking on the orange RSS feed button, and then adding the URL to your feed reader.

Alerting services and RSS feeds are very useful in helping you keep track of current scientific literature in your field. However, the downside is that you can become bombarded with alerts and feeds, and waste rather a lot of time wading through the large amounts of information you receive. To avoid this, be selective in the alerts and RSS feeds you set up. If you sign up to any service and then find it is not particularly useful to you, then unsubscribe from it. Remember, the aim of the tools is to assist you in keeping track of information, not to overwhelm you with unnecessary information.

References

ISI Web of Knowledge (2008). *Web of science search rules* [online]. Available at: http://images.isiknowledge.com/help/WOK/h_database.html (last accessed 1 December 2008).

Open University (2008). *Safari: evaluating information* [online]. Available at: http://www.open.ac.uk/safari (last accessed 1 December 2008).

Chapter 6

Reviewing scientific literature

Introduction

Learning how to review scientific literature competently is an important part of your professional development and is a skill that is fundamental to your work as a scientist. The aim of this chapter is to provide you with a framework for reviewing scientific literature and, in particular, to review original research papers, critically. The guidelines presented here refer to reviewing a research paper *after* it has been published, but apply equally to the review of material prior to its publication.

Specifically, this chapter will:

- provide a brief overview of the aims underpinning the review of scientific literature
- identify the knowledge and skills necessary for conducting a successful review
- present guidelines for reviewing a research paper
- present an annotated example of a written review of a published research paper.

6.1 Why review scientific literature?

As a scientist, you will be expected to review your own work as well as the work of your peers. For example, you may be asked to produce a written or oral critique of a research paper or a grant proposal, or write a literature review, or construct a clear and logical discussion for your own research findings. All of these tasks will involve reviewing scientific literature. This means you will need to assess the strengths and weaknesses of the material you are commenting on and present a well-justified view about the overall quality of the work.

To review critically new research presented in the form of research papers is particularly important, as the material is original and intended to extend existing knowledge in a meaningful way. Consequently, this material must be subjected to careful and rigorous scrutiny if you are to reach a well-justified view about the soundness of the hypotheses or conclusions being proposed in the papers. As described in Chapter 4, this rigorous review takes place at the pre-publication stage to ensure that any omissions, mistakes, and flaws in reasoning are corrected before the research is published. However, peer review does not stop at the point of publication, but continues after the paper has been published. This post-publication review takes place in a variety of ways, including discussions at journal clubs, correspondence with the editors, and citations in other research papers. Publication of research results and the ongoing critical debate is an integral part of the scientific process as it allows independent researchers to repeat the work and either confirm or refute it. This in turn extends and strengthens our understanding of a particular phenomenon. Therefore, ongoing peer review of published material serves a number of valuable purposes which can be summarized as follows:

- It improves the quality of work by identifying any flaws or omissions that were overlooked at the pre-publication stage.
- It allows independent scientists to repeat the work and confirm or refute it.
- It clarifies and enhances our understanding of a particular problem.
- It allows scientists to extend and build on the reported work.
- It helps you to write better yourself.

6.2 The knowledge and skills necessary for reviewing a research paper

All research papers present new research that has not been published previously. Therefore, at the core of each research paper is a primary claim (or claims); that is, the conclusions, hypotheses, or theories which the author is making. This primary claim is based on data which are obtained through a well-designed (or otherwise) experimental strategy formulated to answer specific research questions (or hypotheses). As a reviewer, your role is to assess the strength and weaknesses of the work being reported and to state clearly what must be done to improve the quality of the work. When reviewing a research paper you will typically comment on the following (see Chapter 4, section 4.2.2):

- the soundness of the conclusions being proposed
- the quality of the evidence (i.e. results) being presented in support of the conclusions
- the appropriateness of the experimental strategy used to answer the research questions posed
- the novelty and significance of the findings.

The four points listed below summarize the knowledge and skills necessary to conduct a review successfully.

1. **Have good background knowledge of the subject under review.** How well you are able to review a particular paper will depend on how extensive your knowledge of the topic area is. For example, a key responsibility of a reviewer is to assess

whether the conclusions presented are supported by the experimental results (or whether there are alternative explanations to those being presented by the authors) and what is novel about the work. To make a sensible comment about this, you would need to understand clearly the scientific background to the study and how the work fits into existing published literature. Similarly, you need to be completely familiar with the methodology used, including the statistical methods, to be able to evaluate the quality of the results obtained.

2. **Understand the scientific process**. It is also necessary that you understand the methods and principles that govern the nature of scientific inquiry. A brief outline of the scientific method is presented below for those wanting to refresh their knowledge in this area. There is also a good discussion on certainty and uncertainty in science by Jevning *et al.*, (1994) which is worth reading.

The scientific method is a procedure for testing research questions or hypotheses in a systematic and objective manner. The features of this method are summarized here in five stages:

- **Identifying the topic area to investigate.** The scientific method starts with the researcher identifying a topic to investigate. There are a number of factors which can determine the choice of topic. Some of these are: personal area of interest, the priorities of funding organizations, and the resources available to you such as equipment and technical support.
- **Formulating hypotheses**. Once the topic is selected, the researcher identifies the specific research question he/she will investigate. The research question can be formulated as a series of alternative hypotheses. These hypotheses are predictive statements that are formulated in such a way as to make them empirically testable so that they can be either confirmed or falsified.
- **Testing hypotheses experimentally**. Once the hypotheses have been formulated, the next stage is to devise experiments to exclude or strengthen one or more of the hypotheses. This can be done in two ways; one is through conducting an experiment which may be laboratory, field, or computer based (or a combination of these) in which a particular variable is manipulated by the investigator. Alternateivly, the experiment may be observational, in which the defined variable is not manipulated. Regardless of the nature of the investigation, it must be designed so that it is the strongest test of the hypothesis. This entails the experimenter clearly defining the variable under study and controlling for confounding variables in order to generate meaningful results (Holmes *et al.*, 2006; Ruxton and Colegrave, 2006).
- **Analysis and interpretation of results**. Conclusions are inferred from the results obtained, and the hypothesis is either confirmed or falsified. If falsified, the hypothesis may be modified and then re-tested. If the experiments support the hypothesis, then general explanations can be constructed that partly (or fully) explain the particular observations. Once the work is published, the hypothesis is strengthened through independent researchers repeating the work and building on it (Cassey and Blackburn, 2006; Giles, 2006).

- **Development of scientific theories**. Scientific theories develop from a hypothesis or the synthesis of a group of related hypotheses which have been corroborated through repeated experimentation. Examples of such scientific theories in the biosciences include Darwin's theory of evolution and Mendelian genetics. Theories serve as frameworks into which new discoveries can be integrated. With each new finding a particular theory is strengthened, extended, and fine-tuned. However, on occasions new findings are made that fail to fit into an existing framework despite repeated experimentation. In this case it becomes necessary to replace an existing theory with a new theory. New theories are often initially received with scepticism and it can be some time before they are accepted by the scientific community. An example is the chemiosmotic theory which was first proposed by Paul Mitchell in 1961 and which took at least a further 15 years of contentious debate before it was widely accepted (Prebble and Weber, 2003). This case illustrates a number of key points: one is that it is important that scientists are open-minded enough to consider alternative view points and independent enough to hold well-justified opinions that may be contrary to those in a group. Another is that it highlights the importance of critical debate in moving scientific understanding forward.

3. **Be critical in your appraisal of the information**. Whenever you review scientific literature, whether it is published or unpublished, you should approach it with a critical attitude. To be critical means that you question the information presented and consider the strengths and weakness of the work before formulating a view about its quality. The difference between the uncritical thinker and the critical thinker is that the former merely restates the information contained in a particular piece of work. In contrast, the critical thinker evaluates the information against a set of evaluation criteria and comes to an independent and balanced view about the quality of the work. The critical thinker identifies the strengths and weaknesses of the work and describes alternative ways of conducting and interpreting the work.

This critical approach to scientific research is encapsulated by Pratt (1964) in an article entitled 'Strong inference', in which he writes:

> 'On hearing any scientific explanation or theory put forward ask the question in your mind "but Sir, what experiment could disprove your hypothesis?" and on hearing an experiment being described, ask the mental question "but Sir, what hypothesis does your experiment disprove?"'

When directed at your own work or that of others, these questions will ensure that all alternative explanations for the findings are explored before settling on the most plausible. Do not assume that just because a particular paper has passed the pre-publication peer-review process or because it is published in a high impact factor journal the paper is free of mistakes.

4. **Follow any guidelines provided**. You must adhere to any instructions and guidelines you are supplied with. This includes the questions you need to address, how the review should be structured, and the deadline by which the review should be submitted.

6.3 Reviewing a research paper: a three-stage process

There are three stages to reviewing a research paper. These are:

1. Read through the paper to extract the key points contained within the article. This is necessary to understanding the content of the article.
2. Evaluate the extracted key points against a set of standard evaluation criteria. This is necessary to formulating an independent view about the overall quality of the work.
3. Integrate your observations to present a balanced and well-justified view of the quality of the work. This may be in the form of a written critique or/and an oral critique.

Each of these three stages is described in detail below.

6.3.1 Stage 1: extracting the key points contained within a scientific paper

The first stage in reviewing a research paper is to understand the content. To do this you will need to read through the paper **actively** and extract the key points contained within the paper.

What is reading actively?

Reading actively means that you read for a defined purpose and you select material that is relevant for that purpose. Many strategies have been put forward for reading actively. One is the SQR3 method summarized below. The advantage of reading actively is that you are able to pick out the relevant points more quickly. This is particularly true in the case of research paper articles, as they are commonly structured and presented in an IMRAD format. This means that the paper is organized in sections, with a title and abstract followed by an **I**ntroduction, **M**ethods, **R**esults and **D**iscussion (hence the acronym IMRAD). Most journals use this structure or some a modification of it. For example, the journal *Cell* positions the Methods section at the end of the paper after the Discussion instead of immediately after the Introduction. Other variations to the conventional IMRAD structure are to combine the methods and results under the sub-heading of 'Experimental' or to combine the Results and Discussion sections together under the sub-heading of 'Results and discussion'. Each of the sections contains easily recognizable features regardless of whether the research paper uses the IMRAD structure or a modification of it. These features are summarized in Table 6.1. If you read with anticipation of these features, you will be able to extract the relevant information much more quickly.

TABLE 6.1 The main sections of a research paper together with a brief description of the features associated with each section

Section	Brief description
Title	A short and descriptive title that summarizes the content of the paper
Authors	Names, affiliations, and contact details of people who contributed to the work
Keywords	Words or phrases that are strongly associated with the content of the paper and are used by abstracting and indexing services to index the article
Abstract	A concise summary of the paper outlining the aims, the main methods, the main results, and the main conclusions of the study
Introduction	Provides a clear rationale for the study being undertaken by reviewing the published literature and describing clearly the aims/objectives of the study
Materials and methods	Outlines the materials and methods used to address the aims/objectives stated in the Introduction
Results	A description of the main research findings, illustrated with figures and/or tables
Discussion	Summarizes the main conclusions, critically evaluates the results and compares them with other published studies. Discusses the theoretical/practical implications of the work and makes recommendations for further research
Acknowledgements	Names of people or institutions that have assisted in the work (those not listed as co-authors) including details of funding sources
References	Lists all the references cited in the body of the text
Supplementary information	This information is typically supplied online and includes material that is helpful (but not essential) to the reader in understanding the study

The SQR3 method for reading scientific literature actively

The SQ3R method (described by Rowntree, 1988) is one active reading strategy you may want to utilize for extracting the key points contained within a paper. It involves five stages: survey, question, read, recall, and review (hence the acronym SQR3) and is described below as it would apply to reading a research paper.

- **Survey**. First skim the whole paper to gain an overall impression of it. Identify the main sub-sections; that is, does the paper follow the IMRAD structure or a modification of this? Who are the authors? What is the journal title and publication date of the paper?

- **Question**. The next stage is to construct a series of questions to guide you in selecting material most relevant to your purpose. For the purpose of reviewing a scientific paper, a list of possible questions could be:
  - What are the key objectives (questions) of this study?
  - What are the conclusions?
  - What evidence (i.e. results) supports (or not) the conclusions?
  - What methods were used to obtain the results?
  - What gap in knowledge is this study addressing?
  - What is known at the end of the study that was not known prior to starting the study?
  - Are the finding(s) *important* (i.e. valuable)?
- **Read**. At this stage read through the text to find the answers to the questions. Do not read from beginning to end. A suggested order for reading could be to read the abstract first, then the Introduction, followed by the Results and Discussion sections and then finally the Materials and methods. Look for words or phrases that signal the author's key points (e.g. *we hypothesize, we propose, in contrast to previous work*). Look for the aims/objectives of the study at the end of the Introduction and the conclusions at the beginning or end of the Discussion. At this stage you may want to annotate your paper to highlight key passages and/or points.
- **Recall**. At this phase, you should record the answers to the questions you constructed.
- **Review**. At this final stage, you can check the notes you have made against the research paper to make sure they are accurate and complete. You will need to read the paper more than once and some sections of the paper in more depth than others to understand them clearly.

The SQR3 method is only one strategy for active reading. You may want to consult further textbooks (Buzan, 1994; Northedge *et al.*, 1997; Cottrell, 2005) for additional strategies, or you may already be using a system that works for you! The important thing is to use a method that is appropriate for your purpose and one that suits you.

Extracting the central claim of a scientific paper using argument analysis

A second approach to extracting the key points of a research paper is to identify the central claim made in a paper, i.e. the authors' conclusions. Once you have identified the central claim, then you can move on to identify the evidence (i.e. the results) that are supplied in support of the claim and the corresponding methods used to obtain the results. If you use this approach then you will need to be familiar with the way academic arguments are structured. Some basic principles of argument structure are

are described here. For further reading on argument structure and analysis consult Cottrell (2005).

Academic arguments are made up of two basic components, **reasons** and **conclusions**, where reasons (also called premises) are supplied in order to justify a conclusion made. In science, the reasons supporting a conclusion are usually supplied in the form of factual evidence collated through experimental investigations. A conclusion is then *inferred* from the supporting reasons.

To illustrate the structure of arguments, consider the following two examples.

Example 1

Sexual recognition in mosquitoes (Gibson and Russell, 2006)

When pairs of the same sex mosquitoes are flown together, the wing beat frequencies (and hence flight velocities) of the two individuals diverge. In contrast, when pairs of opposite sex mosquitoes are flown together, their wing beat frequencies converge to the point that they match precisely or very closely. This suggests that mosquitoes change the flight tones created by the beating of their wings to match those of potential mates. The direct outcome of this would be that different flight velocities would serve to spatially separate mosquitoes with different flight velocities but bring together mosquitoes with similar flight velocities allowing mating on the wing.

This argument can be deconstructed into the following two premises, leading to a conclusion as follows:

Premise 1: when pairs of the same sex mosquitoes are flown together, the wing beat frequencies of the two individuals diverge.

Premise 2: when pairs of opposite sex mosquitoes are flown together, their wing beat frequencies converge to the point that that they precisely (or very closely) match in frequency.

Conclusion: mosquitoes change the flight tones created by the beating of their wings to match those of potential mates and the outcome of this may be that mosquitoes with similar flight velocities will be brought spatially close together to allow mating on the wing to take place.

Example 2

Pathological gambling is associated with dopamine agonists used to treat Parkinson's disease

Parkinson's disease (PD) is characterized by a marked reduction in dopamine levels in the nigrostriatal system of the brain as well as in other dopaminergic systems. Treatment is commonly by administration of dopamine precursors or dopamine agonists that restore or improve brain dopaminergic systems. Observational studies have reported an association between dopamine agonist treatment in PD patients and the development of impulsive compulsive gambling in up to

8% of patients studied, the effects of which can be reversed by drug withdrawal (Dodd *et al.*, 2005; Drapier *et al.*, 2006; Pontone *et al.*, 2006). This suggests that pathological gambling may be a side effect of dopamine agonist treatment in PD patients. Dopamine agonists bind to and stimulate the D2-like (D2 or D3) dopamine receptors. Genetic studies have shown that variants of the DRD2 gene which codes for the D2 receptor are associated with gambling behaviour (Noble, 2000). D3 receptors in PD are expressed mainly in the ventral striatal complex, thought to be involved in reward, craving, and emotional processes (Pontone *et al.*, 2006) These studies collectively support the idea that disproportionate stimulation of D2-like receptors by dopamine agonists may be responsible for the pathological gambling observed in dopamine agonist-treated PD patients. It is therefore important that patients are informed of the potential risk of developing compulsive gambling behaviour and to monitor them clinically.

This longer and more complex argument can be broken down into the following premises and conclusion:

Premise 1: reduced levels of dopamine are observed in PD patients, the effects of which are ameliorated by treatment with dopamine agonists.

Premise 2: a series of observational studies have demonstrated a link between pathological gambling and dopamine agonist treatment in up to 8% of patients.

Premise 3: Genetic studies have shown a link between D2-like receptors and repetitive reward-seeking behaviours such as gambling.

Premise 4: D3 receptors are localized in regions of the brain that are involved in reward, craving, and emotional behaviour.

Conclusion: A side effect of treating PD with dopamine agonists is the development of compulsive gambling in patients, which is possibly through disproportionate stimulation of D2-like receptors. It is therefore important that patients are made aware of the possible side-effects and monitored clinically.

The examples presented above illustrate the basic anatomy of arguments. From the examples, you will see that an argument is made up of conclusion(s) and supporting reasons. The number of reasons supplied may vary from paper to paper and you will need to go through the text carefully to make sure you have identified all the reasons and the conclusions.

The principles underpinning argument structure described above can be very useful when applied to your work, in particular when you are attempting to review the work of others or when you are attempting to construct coherent discussions for your own research findings

6.3.2 Stage 2: evaluating the research paper

Once you have extracted the key points of the paper, you will need to evaluate the quality of the work. Evaluation is the main part of the review process, as well as the most demanding. A list of standard evaluation criteria you can use to assist you in the evaluation process are listed in Table 6.2.

TABLE 6.2 Standard evaluation criteria for reviewing a research paper (see also Rangachari and Mierson, 1995; Seals and Tanaka, 2000).

Introduction	Have the authors clearly established a context for their work by identifying what is known and what is unknown about their topic area? Do the authors identify clearly which gap in knowledge they are addressing? Do the authors identify clearly the research questions they are investigating and why it is important to study these questions? Is the proposed work novel?
Materials and methods	Is the experimental strategy utilized the strongest test of the research questions posed or is there an alternative strategy that could be used that would be more appropriate for answering the research questions? Are the experimental methods described in sufficient detail to allow another scientist to repeat the work or at least to evaluate the quality of the findings? Does the experimental strategy control for confounding factors? To answer this question you will need to consider the following: Have the authors used appropriate positive and negative controls? Have the authors used appropriate number of replicates (i.e. sample size)? Have the authors used appropriate sample selection methods? Have the authors used appropriate methods when assigning samples to treatments? Has the experiment been repeated a sufficient number of times? Is each experiment repeated independently of each other? How appropriate are the techniques used for collating the data? Here you will need to consider the level of accuracy and precision of the measuring equipment used to acquire the data. To do this competently you need to understand the theory underlying the experimental technique and limitations of that particular technique. You also need to consider whether there are alternative methods that could produce more precise results and if more than one technique should be used to corroborate the findings How have the data been analysed? Is this the most appropriate way? If any statistical tests are used, are they the most appropriate? If statistical tests have not been used, should they have been?
Results	Have the authors presented replicate data? If yes, how similar are the repeated results? Here you will need to look at the standard deviations or standards errors (if reported) for each data point and compare the variability in results across the different conditions Have the control data been presented? Are the control data clearly differentiated from the experimental data? Have the authors described their findings accurately? Here you should look at the actual data presented rather than reading the authors' account of the differences in the data

TABLE 6.2 Cont'd

	If any mathematical calculations have been performed (e.g. p-values) have these been calculated and interpreted appropriately? Are the tables and figures presented clearly so that the reader can assess the results in relation to the stated aims of the study?
Discussion	Have the authors met the aims of the study that they set out in the Introduction? Are all the conclusions made by the authors adequately supported by the results presented? Is there any alternative explanation for the data presented that the authors have not considered? Do the authors clearly differentiate between fact and speculation? Do the authors speculate excessively? Do the authors discuss their findings in the context of published literature (i.e. studies that support their conclusions as well as those that conflict with their conclusions)? Is there any further information (i.e. additional experiments) required to strengthen the claims being made by the authors? From the discussion is it apparent that the study has addressed a significant gap in our understanding of the problem under investigation?
Additional questions	Are the references accurately reported? Are all the references necessary to understanding the work? Have the authors omitted any significant papers? Are the references sufficiently up to date? Are you satisfied that any disclosed conflicts of interest have not influenced the interpretation of the results? Are you satisfied that the experiments have been conducted in accordance with ethical regulations and requirements (if appropriate)? Is the paper well written and comprehensible to the reader? Does the abstract concisely and accurately summarize the content of the paper without the reader having to refer to the main article?

6.3.3 Stage 3: integrating your observations to produce a written review

At the final stage of the review process you will bring together your observations and notes from the first two stages and integrate these to produce a written review (or in some cases an oral review) of the work. The length and organization of the review will depend on any guidelines you have been supplied with either as part of your programme of study, if you are writing for an assessment, or by the organization for which you are undertaking the review, if you are reviewing for a journal or funding body (the latter you are unlikely to do until the later stages of your career).

Typically a review consists of two parts: a brief summary which comprehensively and accurately summarizes the content of the paper, followed by an evaluation of the quality of the work.

Section 6.4 below presents an example of a written review of a research paper entitled 'The mitochondrial apoptosis-inducing factor plays a role in E2F-1 induced apoptosis in human colon cancer cells'. This paper was published in the *Annals of Surgical Oncology* in 2003 (you can locate a copy of it through Web of Science). Do not worry if you are not familiar with molecular oncology, which is the subject matter of this paper. By reading through the example review (on the left) and the annotations (on the right) you should be able use the content and the comments to direct your line of reasoning to write and organize your own critique for any paper you review.

The example review presented below is divided into four parts:

1. A summary of the work (background, methods, results and conclusions).
2. The strengths of the paper.
3. The weaknesses of the study and suggestions for improvement.
4. Overall opinion and a recommendation either to accept or reject the paper.

The role of the summary is concisely and accurately to present a **brief** overview of the content of the article. This is necessary, as it will inform and orientate readers to the evaluation that follows without their having to go through and read the paper for themselves The evaluation component forms the **main** and **substantive** part of the review. This section focuses on the strengths and weaknesses of the work and highlights any additional studies that are required to support the work further. Overall the review should be written in a constructive manner so that both the strengths and weaknesses are highlighted and any practical suggestions for improving the quality of work are clearly stated.

6.4 Annotated example of a critical review of a published research paper

Research paper: Voburger, S.A., Pataer, A., Yoshida, K., Liu, Y., Lu, X., Swisher, S.G., Hunt, K.K. (2003). The mitochondrial apoptosis-inducing factor plays a role in E2F-1 induced apoptosis in human colon cancer cells. *Annals of Surgical Oncology,* **10**(3), 314–322 (used with kind permission of Springer Science and Business Media).

The example presented here is annotated with comments on the right to assist you in identifying the main features of the review.

Review

Background, methods, results, and conclusions

The aim of this study is to investigate whether apoptotic cell death can be restored by overexpressing E2F-1 in colon cancer cell lines. To achieve this, the authors have successfully transfected four human colon cancer cell lines, the primary cell line SW480 and three metastatic cell lines SW620, HT-29, and KM12L4, with the E2F-1 gene using a replication-deficient adenoviral construct. The results show that E2F-1 overexpression correlates with an increase in the proportion of apoptotic cells and a decrease in the proportion of proliferating cells. This coincides with an upregulation in the survival of the proto-oncogene Bcl-2, activation of the caspases 9 and 3, and release of AIF into the cytoplasmic fraction of the cell. This study also shows that complete inhibition of apoptosis is not achieved when cells are treated with caspase inhibitors. Based on these findings, the authors suggest that E2F-1-induced apoptosis in human colon cancer cell lines occurs through not only a caspase-dependent pathway but also a caspase-independent pathway.

Strengths of the paper

1. This study is novel in that the detailed mechanistic pathway for apoptosis in response to E2F-1 overexpression has not been previously assessed in colon cancer cells. In particular a caspase-independent pathway mediated by E2F-1 has not previously been reported. This study is important as mutations in the pathway confer drug resistance in a subset of cancer patients and hence improved understanding of the pathway could lead to therapeutic strategies that allow cells to overcome resistance.
2. The paper is well written and the figures and tables are clearly presented and easy to follow.
3. The aims of the study are clearly stated and justified in the context of the key published literature in this field.
4. The methodological approach on the whole is sound (but see weaknesses below) and sufficient methodological details are provided so that the work can be reproduced.
5. The results on the whole support the conclusions made by the authors. However, the claim that caspase-independent apoptosis is occurring in colon cancer cells needs substantiating (see weaknesses below).
6. The discussion is well written and the findings are discussed in the context of key relevant published studies without speculating excessively. The findings are integrated and presented in the form of a schematic drawing which describes the proposed apoptotic pathways operating in colon cancer cells and is very useful.

Annotations

The review starts by presenting the overall aims of the study, the main methods, the main results, and the main conclusions. This orientates the reader to the discussion of the strengths and weaknesses that follows

The main strengths of the paper are highlighted in point form. The specific strengths identified for this paper are:

The novelty and significance of the work

The quality and clarity of text and figures and tables

The aims are appropriately justified in the context of the literature

The methods are appropriate (on the whole) and sufficient detail is provided to evaluate or reproduce the claims

Some of the conclusions are adequately supported by the results

The claims are discussed appropriately and fairly in the context of published literature

The weaknesses of the study are identified (again in point form) and specific suggestions are made to improve the quality of work. The particular weaknesses discussed in this example are:

Weaknesses of the paper and suggestions for improvement

The conclusion is based on insufficient evidence and the authors have not considered an alternative explanation for their findings—hence further work is required to substantiate the conclusion

1. The main weakness of this paper is that the authors claim that a caspase-independent apoptotic pathway is operating in colon cancer cells (a finding which has not been previously reported). This conclusion is based on the observation that a 70% decrease (and not full inhibition) in apoptosis is achieved in response to pan-caspase inhibitors. The method used for assessing apoptotic cells is flow cytometric analysis of cells stained with propidium iodide with sub-diploid cells identified (and quantified) as apoptotic (Figure 5). The authors do not consider the possibility that the cells which they claim are undergoing caspase-independent death are in fact necrotic cells and therefore there is no caspase-independent apoptosis taking place. PI does not conclusively differentiate between apoptotic and necrotic cells and therefore an additional quantitative apoptotic test such as Annexin V or TUNEL should be conducted to substantiate the claim the authors are making.

Incompleteness of the data presented and how this could affect the way the results are interpreted is noted

2. Figure 2B (table)—The authors do not provide complete data on the status of pRb, p14Arf, and Cyclin D1 in some of the cell lines. The status could be rapidly identified through immunohistochemistry using specific antibodies (which are all widely available) directed towards the relevant antigen. This is important as the status of any of these proteins could influence E2F-1 overexpression and hence how the results are interpreted.

The review concludes with a brief paragraph that summarizes the overall view of the reviewer and makes a recommendation to the editor to accept the paper but with modifications

Overall opinion and recommendation to accept/reject

The study is valuable in beginning to characterize the E2F-1-induced apoptotic pathway with the view to utilizing our understanding of the mechanism to treat a sub-set of cancer patients. However, the suggestion that caspase-independent apoptosis is taking place needs to be substantiated through quantitative techniques such as Annexin V or TUNEL.

Recommendation: accept with modifications

References

Buzan, T. (1994). *The mind mapping book*. BBC Publications, London.

Cottrell, S. (2005). *Critical thinking skills: developing effective analysis and argument*. Palgrave Macmillan, Basingstoke.

Cassey, P. and Blackburn, T. M. (2006). Reproducibility and repeatability in ecology. *Bioscience*, **56**(12), 958–959.

Dodd, M.L., Klos, K.J., Bower, J.H., Geda, Y.E., Josephs, K.A., Ahlskog, E.J. (2005). Pathological gambling caused by drugs used to treat Parkinson's disease. *Arch. Neurol.*, **62**, 1377–1381.

Drapier, D., Drapier, S., Sauleau, P., Derkinderen, P., Damier, P., Allain, H., Verin, M., Millet, B. (2006). Pathological gambling secondary to dopaminergic therapy in Parkinson's disease. *Psychiatry. Res.*, **144**, 241–244.

Gibson, G. and Russell, I. (2006). Flying in tune: sexual recognition in mosquitoes. *Curr. Biol.*, **11**, 1311–1316.

Giles, G. (2006). The trouble with replication. *Nature*, **442**, 344–347.

Holmes, D., Moody P., Dine, D. (2006). *Research methods for the biosciences.* Oxford University Press, Oxford.

Jevning, R., Anand, R., Bedebach, M. (1994). Certainty and uncertainty in science: the subjective concept of probability in physiology and medicine. *Adv. Physiol. Educ.*, **12**(1), S113–S119.

Noble, E.P. (2000). The DRD2 gene in psychiatric and neurological disorders and its phenotypes. *Pharmacogenomics*, **3**, 309–333.

Northedge, A., Thomas, J., Lane, A., Peasgood, A. (1997). *The sciences good study guide.* Open University, Milton Keynes.

Pratt, J.R. (1964). Strong inference. *Science*, **146**(3642), 347–353.

Prebble, J. and Weber, B. (2003). *Wandering in the gardens of the mind: Peter Mitchell and the making of Glynn.* Oxford University Press, Oxford.

Pontone, G., Williams, J.R., Bassett, S.S., Marsh, L. (2006). Clinical features associated with impulse control disorders in Parkinson disease. *Neurology*, **67**, 1258–1261.

Rangachari, P.K. and Mierson, S. (1995). A checklist to help students analyse published articles in basic medical sciences. *Adv. Physiol. Educ.*, **13**(1), S21–S25.

Ruxton, G.D. and Colegrave, N. (2006). *Experimental design for the life sciences*, 2nd edn. Oxford University Press, Oxford.

Rowntree, D. (1988). *Learn how to study: a guide for students of all ages*, 3rd edn. Macmillan, London.

Seals, D. R. and Tanaka, H. (2000). Manuscript peer review: a helpful checklist for students and novice referees. *Adv. Physiol. Educ.*, **23**(1), 52–58.

Chapter 7

Writing a literature review

➲ Introduction

The literature review provides a critical overview of recent research in a particular topic area and can be written either as a self-contained report or as an introduction to a larger piece of work. Undertaking a literature review has many benefits, ranging from expanding your knowledge of a specific topic to enhancing your information retrieval and critical review skills. The aim of this chapter is to develop your ability to write well-structured literature reviews.

Specifically, this chapter will:

- identify the different types and purposes of literature reviews
- describe a six-stage process for writing a literature review
- present an annotated example of a completed literature review
- provide a checklist for reviewing drafts of literature reviews.

7.1 What is a literature review?

Literature reviews are secondary sources of information (see Chapter 4, section 4.1) and therefore do not report new primary data. Instead they attempt to bring together primary research findings from different publications. These publications can include journal articles, books, conference proceedings, and theses relevant to a particular topic.

Literature reviews can be viewed as being of two types. They can be either self-contained reports of the type found in published journals on a particular topic (see Chapter 4, Table 4.1) or as an introduction to, and justification for, engaging in primary research (Bruce, 1994). The latter category includes:

- A review as part of a research proposal (Chapter 8).
- An introduction to a research paper reporting new work (Chapter 9).
- A chapter in a dissertation or thesis (Chapter 12).

In general, self-contained literature reviews can either be broad in scope or specialized. For example, if the aim of a literature review is to provide a general introduction to a topic area, then it is likely to be broad in scope and to cover a wide number of related topics but at a less detailed level. If its aim is to provide a specialist review of a particular topic area, then the number of sub-topics covered may be restricted, but the level of detail will be substantially higher.

When literature reviews form part of a larger piece of work such as a research proposal, a dissertation, or an introduction to a research paper, the scope of the material covered is restricted to studies that are directly relevant to the specific question(s) addressed by the new study. The purpose of the review in these instances is to present the rationale on which the research aims and hypotheses rest.

Regardless of whether literature reviews are written as self-contained reports or as introductions to embarking on a research problem, they all have the following features in common:

- They synthesize findings from different reports into a summary of the key facts, concepts, and theories that make up the topic area.
- They identify areas of controversy in the literature and may attempt to explain the differences between studies.
- They identify gaps in the work and formulate meaningful questions that require further research.
- Overall, the review establishes connections between the different studies and provides an integrated and new perspective on the topic.

Undertaking a literature review has many benefits, including the following (adapted from Bourner, 1996):

- You will acquire knowledge of the facts, theories, and concepts that make up the subject area and become familiar with the main methods and techniques that are commonly used in the field of work.
- You will be able to identify areas of current interest and gaps in the work which still need addressing.
- You will be able to identify seminal works and the researchers who are active in the field.
- You will be able to establish a context for your own work.
- You will enhance your information-seeking and retrieval skills and substantially improve your ability to understand and critically review scientific information.
- You will develop your ability to write well-structured reports and present clear and coherent arguments.

7.2 Writing a literature review—a six-stage process

Reviewing scientific literature includes locating information that is relevant to the topic area, evaluating the material, and then synthesizing it to produce a written review on the topic area. The process of writing the review can thus be divided into the following six steps (adapted from Cooper, 1982):

1. Define the purpose for which you are reviewing the literature.
2. Define the topic area that you are working on as precisely as possible.
3. Conduct a search of the literature to locate information relevant to the question.
4. Evaluate the literature located.
5. Synthesize the information.
6. Write the literature review.

The next section describes each of these stages in turn.

7.2.1 Stage 1: defining the purpose for reviewing the literature

The first step in the literature review process is to define clearly the purpose for reviewing the literature. To assist you, use the following questions:

- **What type of review is it?** This could be a self-contained review or an introduction to a dissertation, thesis, research paper, or research proposal.
- **What is the length of the review?** Some reviews will be short, such as introductions to a research paper and brief self-contained literature reviews, and others may be longer, such as an introduction to a dissertation or thesis.
- **How broad should the coverage of the material be?** This will depend on the purpose of the review. If it is an introduction to a Master's dissertation or a PhD thesis, then you will need to review a large body of work. If it is an introduction to a paper reporting original research or as part of a research proposal, the review will be highly focused and the studies restricted to those that directly relate to your research question.

7.2.2 Stage 2: defining the topic

Once the purpose is defined, then move on to the second stage, which is to define as precisely as possible the question or problem you are researching. Defining the topic will narrow the scope of your review and serve as a starting point for locating the most relevant sources of information.

To illustrate how a topic can be defined, consider the situation where you are asked to write a self-contained literature review on the following topic:

The molecular basis of treatment failures in *Chlamydia trachomatis* infections

(A short annotated literature review for this topic is included in section 7.3.)

First you could identify the key terms central to the topic by circling them (Figure 7.1).

FIGURE 7.1 Highlighting key terms.

Following on from this you can begin to map out the content by brainstorming (see Chapter 1, Table 1.2) a series of possible concepts for inclusion in the review. This is illustrated in the form of a brainstorming map in Figure 7.2.

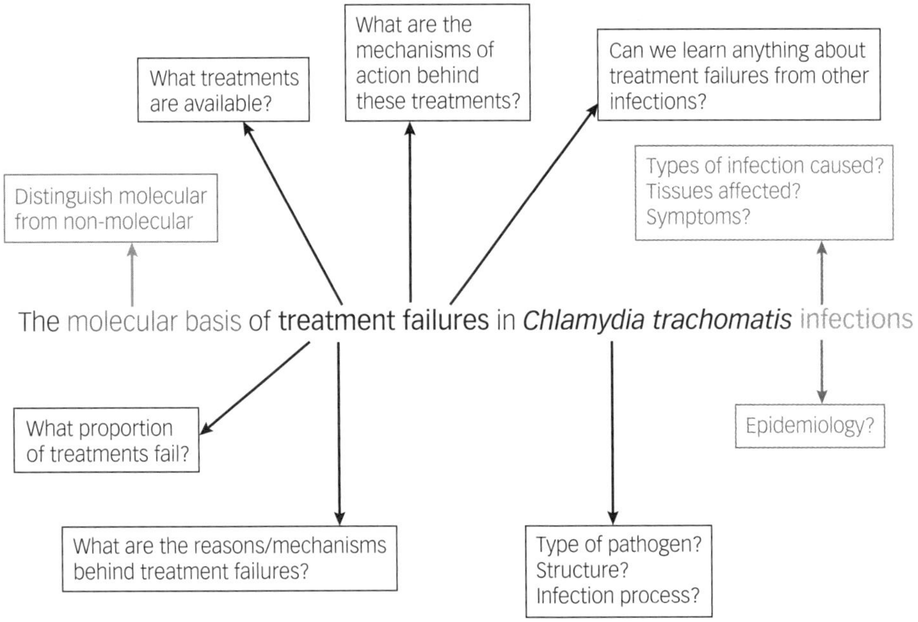

FIGURE 7.2 Brainstorm of possible concepts that may be relevant for inclusion in the literature review.

7.2.3 Stage 3: conducting a literature search

Once you have analysed the topic title and brainstormed possible concepts to include in the review, then move on to search the literature. A strategy for conducting a systematic literature search was outlined in Chapter 5 (section 5.4) and is summarized below:

- Start by reading one or two current review articles to obtain an overview of the facts and theories in the topic area.

- Use the 'ancestry approach' to literature searches. This approach retrieves information by tracking citations from one paper to another. For example, if you have a particularly good reference then use its reference list to track down further suitable references and from these papers track down further references, and so on.
- Use the citation search facility to locate new articles that cite an older reference. This is an excellent way of widening your search in a particular topic area. Some databases such as PubMed have a 'related articles' facility, which links you through to articles related to your reference, and this is also a very useful way of finding relevant articles and widening your search.
- Do not confine your searches to a single database. Instead, carry out searches of multiple databases. This way you avoid missing relevant studies.
- Follow up broad preliminary searches with further searches which are narrower and much more defined. During your initial searches you will be exploring the full extent of the literature that exists in the field and hence these searches will be relatively broad in scope. As you conduct further searches, these should become narrower and much more specific, addressing particular components of the review question. For example, in our current review topic, you may initially conduct a search of the different types of mechanisms underpinning treatment failures in patients infected with *Chlamydia trachomatis* and then follow this up by narrower searches focusing on each type of mechanism (e.g. persistent state, classical antibiotic resistance).
- Each literature search will yield a list of citations, but not all of these may be relevant to your topic. You will therefore need to make a judgement about which papers are related to your subject and therefore useful to read in full. Reading the title and abstract of each citation will usually give you a good indication of the relevance of the article's content in relation to your topic area. A high number of citations for a particular article is also a good indicator of an interesting article, and such references are also worth looking into.

7.2.4 **Stage 4: evaluating the literature**

Once you have identified a set of preliminary papers, then read through the content critically to understand and extract the key points of the article (how to read scientific literature critically was discussed in Chapter 6). The following is a summary of how to read and extract information for the purpose of writing a literature review.

1. First, organize your papers into categories (based on titles, abstracts, or keywords) so that papers addressing similar topics are grouped together. Try and read the articles in the same category at about the same time. This will reinforce your understanding of the material and help you to make connections between the various studies. It will also assist you in compiling tables (e.g. Table 7.1) in which you summarize the main content of the articles.
2. Read the material actively (see Chapter 6, section 6.3.1). This means being clear from the outset what types of information you are aiming to extract from the

literature and then seeking out the information when reading the paper. For the purpose of a review, the types of information to extract could include some (or all) of the following (adapted from Hart, 2005):

- key facts, theories, and concepts that define the subject area
- definitions of key terms
- main methods and techniques used
- any questions that remain unanswered
- timeline of developments in the subject area—in particular, emerging concepts
- areas of controversy in the literature
- wider implications of the work
- relationships between various studies (e.g. how they compare and contrast, how they support and conflict with each other).

3. Have a clear system for making notes. There are a number of different methods that can be used for effective note-making, including highlighting the text, annotating in the margins, and constructing summary tables of the type described in stage 5 below (section 7.2.5).

7.2.5 Stage 5: synthesizing the information

By the end of stage 4 you will have a selection of annotated articles as well as a series of notes containing information that may be appropriate for inclusion in your literature review. The next step is to organize your notes in a way that will help you make connections between the various studies.

A simple but effective method for organizing notes is to construct tables such as Table 7.1 in which you tabulate the key information from each study.

TABLE 7.1 Example of a table summarizing key information from primary research publications

Authors/ publication date*	Key hypothesis/ research questions investigated	Key methods used	Key findings	Key conclusions	Additional information (e.g. type of study; *in vivo*, *in vitro*, human, animal, etc.)

* If you have more than one paper from the same author(s) in the same year add the suffix 'a', 'b', etc. (e.g. 2006a, 2006b) and make sure you append the suffix to the full reference in your database of references *at this stage*.

Summary tables such as Table 7.1 are very useful in helping to identify the main trends in the literature and the relationships between the various studies, and therefore this method is sometimes termed a **meta-summary** (Aveyard, 2007). If you are reviewing a large body of literature, you may consider building up multiple summary tables, with each table summarizing a different aspect of the topic area. For instance, one table may summarize studies that deal with one particular theory while a second summarizes studies dealing with an alternative theory. Alternatively, one table may focus on *in vivo* studies while a second focuses on *in vitro* studies. The way you group together and summarize the literature will depend on the content of your review. Overall, you should aim to build up tables in a way that will help you to make comparisons between studies and identify reasons for the similarities and differences in findings.

Mind maps are an alternative (or additional) method that can used to obtain an overview of a topic and identify relationships between the various studies. In a mind map the sub-topics are arranged radially around the central topic (see Chapter 1, Table 1.2). A mind map for our example literature review topic is included in Figure 7.3. Here the central topic 'The molecular basis for treatment failures in *Chlamydia trachomatis* infections' is written in a box in the centre of the page. Then working in a clockwise direction, the sub-topics are written, again in boxes, branching from the centre. Further sub-points then branch off from the main points.

7.2.6 Stage 6: writing the literature review

As described previously (section 7.1), when writing a literature review you are aiming to present the current state of knowledge in a particular field by integrating information from different primary publications.

Therefore, a good literature review will:

- Demonstrate that the author has reviewed a sufficient body of literature to provide a *comprehensive* overview of the most *current* thinking in the topic area. This means that the references cited should be wide ranging, relevant, and as up to date as possible.
- Show evidence of critical thinking. This means considering competing opinions that exist in the field and discussing the extent to which the literature supports each of them. You should describe the similarities and differences observed between various findings and attempt to provide explanations for these, based on empirical evidence.
- Provide a new perspective of the work. This means you will need to make connections between the different studies and integrate the information in a way which sheds new light on the topic area. This includes discussing any emerging concepts in the field, identifying gaps, and formulating hypotheses and questions that require further study.
- Not merely list facts and studies. Instead, the review should integrate the concepts to provide a coherent and sustained analysis of the main themes in the topic area.

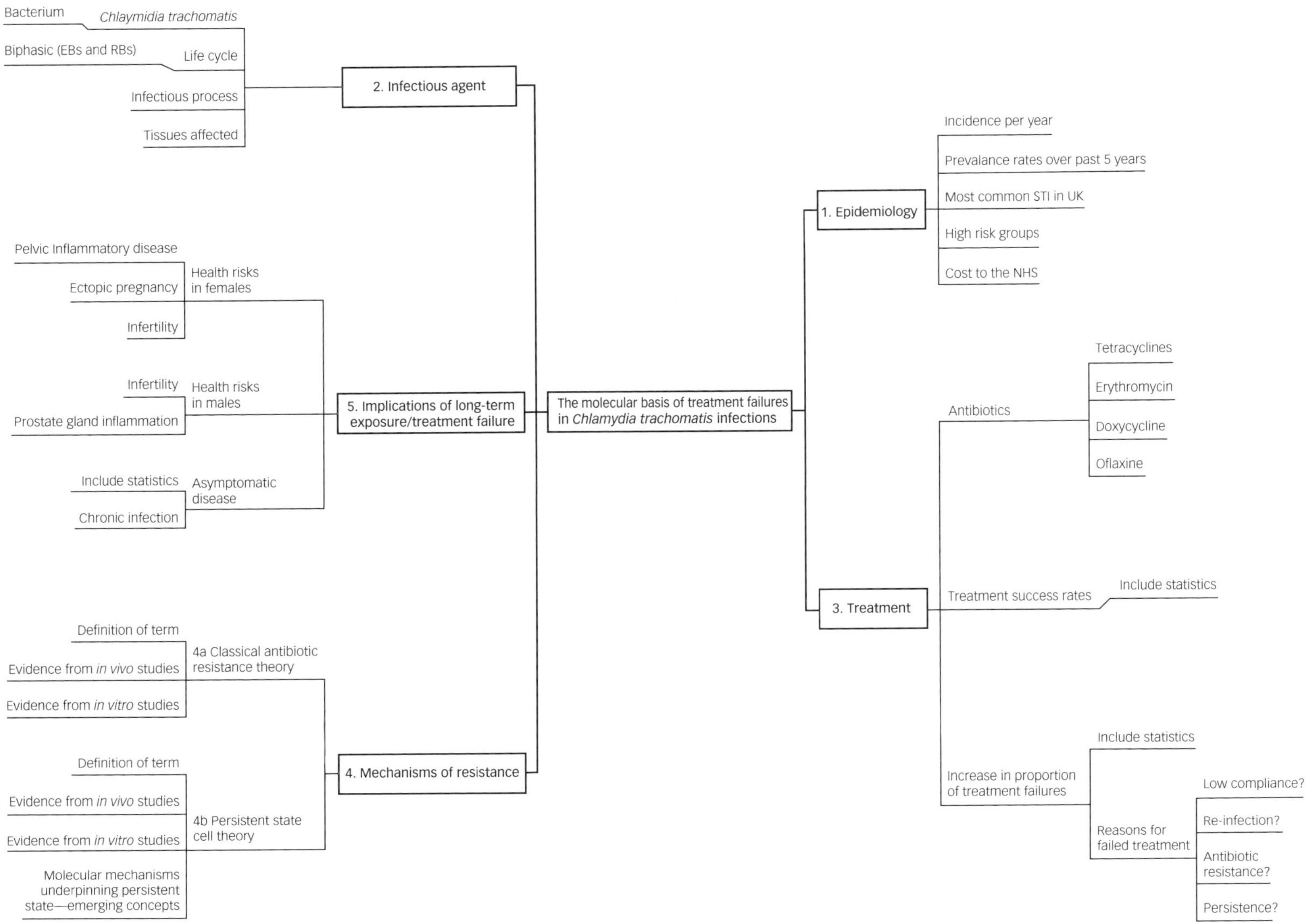

FIGURE 7.3 Example of a mind map for the topic 'The molecular basis of treatment failures in *Chlamydia trachomatis* infections'.

Structure of the review

Literature reviews are structured into five main parts: title, abstract, introduction, the main body of the review, conclusion, and reference list. A description of each of these components is summarized in Table 7.2.

TABLE 7.2 The main components of a literature review

Component	Description
Title	Concisely describes the subject of the review
Abstract	Should convey the key content of the article to the reader in a concise manner. Usually 200–300 words in length. Should not include illustrations or references. Abbreviations should be kept to an absolute minimum (see Chapter 10, section 10.1)
Introduction	Approximately 10% of the report. Aims to orientate the reader to the topic under review and achieves this by establishing clearly the scope of topic under review and explaining why it is important to review the topic
Main body	The main body of the review accounts for the majority of the article and aims to integrate the different studies in the topic area to produce a new perspective of the work. It achieves this by: • summarizing the key facts, concepts, and theories that make up the topic area • considering competing views that exist in the field and discussing the extent to which the literature supports each of them • identifying emerging concepts in the field • identifying gaps in the literature. It should be divided into sub-sections with appropriate sub-headings.
Conclusion	The aim of the conclusion is to sum up the key ideas discussed in the main body of the review and provide a clear direction for future work. It achieves this by: • summarizing the main ideas discussed in the review, including any inconsistencies and gaps identified in the literature • suggesting wider implications of the work • proposing meaningful questions for further research. In the case of a review which is written as an introduction to and justification for engaging in primary research, then the concluding paragraph will identify the key questions the research will address
Reference list	Includes all the references cited in the review and is placed at the end of the document. The references cited should demonstrate that the work is based on authoritative sources of information and a thorough and up-to-date review of the literature has been conducted. The references should be cited accurately and use a consistent referencing style (see Chapter 1, section 1.2).

The writing process

To write a well-structured, concise, and coherent review, you should give yourself **sufficient time** to **plan**, **write**, and **revise** your work. These three stages of writing were described in Chapter 1 (section 1.3) and have summarized below.

Planning your writing.

The first step in writing the literature review is to plan the content and the order in which to introduce the material.

1. **Which information should I include?** Undoubtedly, you will have more information than can be included within the word limit stipulated. Therefore, you will need to make a judgement on which literature is important enough to include. To help you decide, consider the relevance of the information in relation to your topic, the importance of the reference within the body of published literature and how current the reference is. Once the information is selected, then consider the comparative importance of each study. The more important a study is the more space you can devote to it in the review.
2. **In which order should I present the information?** Information in a review is usually arranged thematically. The annotated example of the review included in section 7.3 uses this arrangement. However, a chronological arrangement is also suitable if you are, for example, writing an article that presents a timeline of developments.

There are two techniques that can be used to facilitate the planning stages of your work. One is the **mind map** and the second is the **linear plan**. In the case of the mind map, first produce a brainstorm of the main points and sub-points (as we did in Figure 7.3) and then number the points on the map to indicate the sequence in which to introduce the ideas. In the case of the linear plan (Box 7.1), it is divided into the three main sections of the review: the introduction, the main body, and the conclusion. Under each section the main themes are presented in a sequential order.

The form of planning you use is a matter of personal preference, but it is worth spending some time on the planning stage to ensure that key points are not omitted and that the writing is organized in a logical and coherent manner. This will make your work easier to follow and therefore more comprehensible to the reader.

Writing a first draft

When writing a first draft, you do not need to write from start to finish. Instead start with any section that seems the easiest. You can also add sub-headings and then add details below each sub-heading. This can facilitate the writing process, and any sub-headings that you do not require can be removed later during the redrafting stage. The aim of the first draft is to produce a tentative outline of your work which can be refined and fleshed out with details as you progress through your writing. Almost certainly you will find gaps at this stage (or require further clarification of certain points) and will therefore need to refine your literature search further. If after an intensive search you find nothing else, then note in your review that there is a gap in our knowledge of the subject.

BOX 7.1 An example of a linear plan for the literature review 'The molecular basis for treatment failures in *Chlamydia trachomatis* infections'

Introduction (1 paragraph)

- Outline the main content of the review: mechanisms underpinning treatment failures (antibiotic resistance and persistent state cell theory)
- Establish why the topic is important (e.g. increasing prevalence, increased treatment failures, infections, and associated health risks)

Main body (5 sub-topics—approximately 8 paragraphs)

1. Epidemiology and pathogenesis of the disease
    - Causative organism
    - Most common STI in the UK
    - Increasing prevalence in the UK
    - High risk groups
    - Cost to the NHS to treat infections
    - Life cycle of *Chlamydia trachomatis* and survival in host cell
2. Treatment
    - By antibiotics
    - Treatment success rates and failure rates (include statistics)
    - Possible causes of increased failure rates (not low compliance and reinfection but ability of cells to survive treatment)
    - Exact mechanisms unclear—but two theories proposed: (1) classical antibiotic resistance theory and (2) persistent state cell theory
3. Classical antibiotic resistance theory
    - Definition of term—antibiotic resistance
    - Studies that show survival through antibiotic resistance: *in vitro*
    - Studies that show survival through antibiotic resistance: *in vivo*
4. Persistent state cell theory
    - Definition of term—persistent state cell
    - Studies that show survival through entering a persistent state: *in vitro*
    - Studies that show survival through entering a persistent state: *in vivo*
    - Molecular mechanisms underpinning persistent state (current understanding and emerging concepts)
5. Implications of long-term exposure/treatment failure to *Chlamydia trachomatis*
    - Health risks in males and females
    - Asymptomatic (include statistics)

Conclusion (1 paragraph)

- Summarize the main points of the review and establish a direction for further work: further *in vivo* studies required on how persistence develops and underlying molecular mechanisms in order to improve treatment strategies

When writing your literature review:

- Make sure your work is based on authoritative sources of information (Chapter 1, section 1.2) and your review conveys the content of the sources accurately.
- Write concisely and clearly. Use short sentences and avoid needless repetition. Try and use simple words as far as possible and avoid unnecessary technical jargon. Technical words are sometimes necessary, of course, but on the whole avoid long and complex terms. Also avoid the use of bullet points and lists. Overall your text should flow as a continuous piece of writing.
- Use sub-headings to structure your work. This will make it easier to read and to follow.
- Consider using illustrations (figures and/or tables) in your review. When used appropriately, illustrations can be very useful in explaining difficult concepts and helping the reader to see relationships between different components. Remember, if you decide to use published illustrations for work that is intended for publication, then you will need to obtain permission from the copyright owner (see Chapter 3, section 3.6).
- Make clear which is your own work and which is the work of others by referencing your work accurately and completely. This includes citing the reference within the text at the point you refer to it and listing fully all the sources that you have cited in-text at the end of your document in the form of a reference list (see Chapter 3, section 3.4).

Revising your draft

Be prepared to rewrite your first draft several times and use the checklist provided in Box 7.2 to help you refine the drafts. As you refine each successive draft, it will help if you focus on different elements of your writing. For example, the first time you redraft you may focus on the content and rearrange, delete, or add further text. Subsequently you may focus on checking the accuracy of the references. On further readings you may decide to focus on the spelling, grammar, and punctuation. With each successive draft your writing will improve and you will end with a professional final version that is ready for submission.

BOX 7.2 Reviewing drafts of the literature review—checklist

General structure

- ❑ Does the literature review include a title that is brief and descriptive of the content of the review?
- ❑ Is the literature review structured into three sections; the introduction, the main body of the text, and the conclusion?
- ❑ Is an abstract required as part of the review? If yes, is it included?

BOX 7.2 Cont'd

Content

Introduction

- ❑ Does the introduction establish clearly why it is important to review the topic?
- ❑ Does the introduction provide an overview of the key points in the main body of the literature?

Main body of the text

- ❑ Does the main body of the text accurately summarize the key facts and theories in the literature?
- ❑ Does the main body of the text identify and attempt to explain similarities and differences between various studies?
- ❑ Does the main body of the text identify emerging concepts in the field?
- ❑ Does the main body of the text identify any gaps in the literature?
- ❑ Is all the material used in the main body relevant to the topic under review?

Conclusion

- ❑ Does the conclusion summarize the key points covered in the main body of the text?
- ❑ Does the conclusion suggest wider implications of the work?
- ❑ Does the conclusion formulate questions that need further research?
- ❑ If the review is an introduction to and justification for engaging in primary research, are the key objectives of the study clearly articulated?

References

- ❑ Are your sources of information authoritative?
- ❑ Are your references relevant and up to date?
- ❑ Are all the sources you have used appropriately acknowledged in the text?
- ❑ Are all the sources cited in the text also listed in the reference list?
- ❑ Have you used a consistent style of referencing both in the text and in compiling the reference list?
- ❑ Does the format of the references conform to any guidelines provided?
- ❑ Have you obtained permission for any tables or figures you have reproduced for publication purposes?

Organization and coherence of article

- ❑ Have you used appropriate sub-headings to structure the review?
- ❑ Do the paragraphs follow in a logical manner so that the line of reasoning is clear?
- ❑ Does each paragraph cover one topic only?
- ❑ Are appropriate transitions used to link each successive paragraph?

- ❑ Are there any unclear or long sentences that can be rewritten to improve clarity?
- ❑ Are there any places where you have repeated what you have said elsewhere and therefore the content needs deleting?
- ❑ If you have used illustrations, are they cited in the text?
- ❑ Does each illustration include a sequential number, a title, and a legend?
- ❑ If you have used any abbreviations, symbols, or units are they in the correct format and consistent in style?
- ❑ Have you defined abbreviations at first use?
- ❑ If the review is long, have you summarized key points at intermediate places in the text to help the reader follow your line of reasoning?
- ❑ Have you checked for mistakes in spelling, grammar, and punctuation?
- ❑ Have you checked the layout of the document (e.g. font and size, spacing, margin size) and the length to see that it conforms to any guidelines provided?

7.3 Annotated example of a short literature review

This review is an example of a short self-contained report on the topic 'The molecular basis of treatment failures in *Chlamydia trachomatis* infections'. The annotations on the right describe the key features of the literature review including the perspective and organization of the content. Regardless of whether or not you are familiar with the subject matter of this review, by reading through the example and the associated annotations, you should be able to apply the same overarching principles described here to the organization and content of any literature review you write.

Literature review

The molecular basis of treatment failures in *Chlamydia trachomatis* infections

The infection rates of the sexually transmitted disease chlamydia caused by the bacterium *Chlamydia trachomatis* are rising in the UK, particularly amongst the 16–24 year old age group. Although the majority of infections can be successfully treated with a course of antibiotics, there are a growing number of infections that fail to be cleared by this

Annotations

Title: brief and concise, describing the topic under review

Introduction: provides an overview of the main trends in the literature and hence clearly defines the scope of the review. It establishes the reasons why the topic under review is important by highlighting the extent of the problem and the gaps in the knowledge base

> The last sentence in the introduction clearly identifies the perspective taken by the author

treatment. Whilst the exact mechanisms underpinning treatment failures are unclear, it is considered to be a consequence of either antibiotic resistance or the ability of cells to enter a persistent state. In both cases, the outcome is the same: recurrent chlamydial infection. Long-lasting infection can pose significant health risks, in particular to women, who can become infertile. In light of these factors, this paper proposes a case for investigating further the exact mechanisms by which *Chlamydia trachomatis* survive antibiotic treatment.

> **Main body of the review:** This is divided into eight paragraphs and organized *thematically* so that studies relating to a particular topic area are grouped and discussed together. You will note that the amount of detail supplied for each study differs depending on its comparative importance in the literature and on the space available (i.e. length of review). The first paragraph focuses on the epidemiology of the disease and establishes the rising trend in infection rates and increased rates of treatment failure

Chlamydia trachomatis is an obligate intracellular bacterium which causes infection of the male and female genital tract. It is currently the most common sexually transmitted infection (STI) in the UK—genitourinary medicine (GUM) clinics diagnosed 182.6 new chlamydial infection cases (per 100 000 population) in 2005. Over the past 5 years the diagnosis rates of chlamydial infections have increased by 35%, which is significantly higher than any other STI. Rates of diagnosis are highest amongst young sexually active people, in particular amongst women aged 16–19 years and men aged 20–24 years. The cost to the UK of treating chlamydia infections is substantial, totalling around £100 million per year (Health Protection Agency, 2006).

> The second paragraph discusses the life cycle of *Chlamydia trachomatis* and relates this to the infectious process. This is important as treatment failures are linked to the cycle and therefore treatment strategies will attempt to subvert the cycle.
>
> Note: abbreviations are defined at first use

Chlamydia trachomatis is able to cause disease by utilizing a unique biphasic life cycle in which the organism alternates between an intracellular, non-infective (but metabolically active) reticulate body (RB) and an extra-cellular, infectious (and metabolically inactive) elementary body (EB). The cycle begins with an EB attaching to the host cell membrane and entering the cell by stimulating their endocytosis into a host-derived vacuole. Here they differentiate into the larger RB. These RBs then multiply via binary fission and differentiate back into EBs. These newly formed EBS are then released from the host cell, either by lysis or by exocytosis, to begin a new cycle of infection (Binet and Maurelli, 2005).

> The third paragraph moves on to discuss treatment and links this to increased rates of treatment failures. The possibility that the increased infection rates could be due to low compliance and re-infection are eliminated by identifying supporting studies

Chlamydial infections are treated using antibiotics such as doxycycline, tetracycline, azithromycin, and oflaxine. Depending on the antibiotic administered, cure rates of >95% (measured microbiologically) have been demonstrated 2–5 weeks post-treatment (Lau and Qureshi, 2002). However, recent studies have reported that up to 15% of antibiotic-treated patients fail to be cured (La Montagne *et al.*, 2004; Whittington *et al.*, 2001) and that factors other than re-infection and low compliance with the drug-treatment regime are responsible for the observed treatment failure (Mpiga and Ravaoarinora, 2006; Department of Health, 2002).

Currently the exact mechanisms underpinning treatment failures are unclear but are considered to be a consequence of either classical antibiotic resistance or the ability of cells to enter a persistent state. In both cases a proportion of cells survive treatment and can re-establish an infection at a later stage.

In classical antibiotic resistance, the bacterium alters the target of the antibiotic via a mutation in the gene coding for the target so that the efficacy of the antibiotic is reduced and the organism can persist within the host cell. The persistence of antibiotic resistance cells in response to sub-optimal antibiotic treatment has been demonstrated extensively *in vitro* for fluroquinolones (Dessus-Babus *et al.*, 1998), rifampicin (Suchland *et al.*, 2003a), tetracycline and its derivatives (Jones *et al.*, 1990), and the macrolides (Misyurina *et al.*, 2004). However, only a limited number of studies have demonstrated the same *in vivo* (Bragina *et al.*, 2001; Dean *et al.*, 2000; Somani *et al.*, 2000; Lefevre and Lepargneur, 1998). In addition, all of these *in vivo* studies report that the bacteria appear in a morphologically aberrant state under microscopic analysis, indicating that they are more likely to be in a persistent state (as in Bragina *et al.*, 2001) as opposed to classical resistance (in which bacteria continue to undergo a normal cell cycle) even in the presence of the antibiotic.

In the case of the persistent state cell theory, it is proposed that under stressful conditions the organism enters an intracellular, metabolically less active, morphologically aberrant state that makes it less sensitive to antibiotic treatment. A number of *in vitro* studies have shown that different factors such as sub-inhibitory levels of the cytokine interferon-γ (as can be found in the human host following a T-helper 1 cell response to the infection) (Pantoja *et al.*, 2001), removal of tryptophan (an amino acid that *Chlamydia* takes from the host cell in order to grow) (Coles *et al.*, 1993), and, importantly, sub-inhibitory concentrations of antibiotics (Johnson and Hobson, 1977; Segreti *et al.*, 1992; Dreses-Werringloer *et al.*, 2000) are all able to induce persistent state *Chlamydia trachomatis* cells which can be reactivated following removal of the stressful condition. The presence of morphologically aberrant forms of *Chlamydia trachomatis in vivo* has also been reported following antibiotic treatment for chronic arthritis (Reiter's syndrome) (Nanagara *et al.*, 1995) and in a limited and inconclusive study in humans (Bragina *et al.*, 2001), and also in the various reports of antibiotic resistance as discussed under classical resistance above. However, it is not known how long these cells can persist in a dormant state *in vivo* and whether removal of the antibiotic can reactivate the organism (Suchland *et al.*, 2003b; Somani *et al.*, 2000).

> The fourth paragraph moves on to propose two mechanisms underpinning treatment failure, classical antibiotic resistance and persistent state cell theory. *Note that the author clearly states that the exact mechanisms at this stage are not clear*

> The next paragraph moves on to describe studies that support the classical antibiotic resistance theory. These studies are grouped so that *in vitro* studies are discussed first followed by *in vivo* studies. The difference in the extent of the literature between *in vitro* and *in vivo* studies is noted (*..only a limited number of studies have demonstrated the same in vivo . . .*). In addition, unrepresentative findings are highlighted and an explanation provided for the unrepresentative observation. Overall, the paragraph establishes that the evidence from *in vivo* studies supporting the classical antibiotic resistance theory is limited

> The review then moves on to discuss studies supporting the second theory for treatment failures. The paragraph is organized as previously so that *in vitro* studies are discussed first, followed by *in vivo* studies. This organization makes it much easier for the reader to follow the line of reasoning employed by the author. Again gaps in knowledge base are clearly identified

Having established that the evidence for persistent state cells appears stronger than for the classical antibiotic resistance theory, the author moves on to describe in detail the possible molecular mechanisms underpinning the persistent state cell. This paragraph establishes that this is an emerging area of study and that much more work in this area is required to understand the precise mechanistic details underlying the persistent state

Studies investigating the molecular mechanism underpinning the persistent state cell indicate that deprivation of essential acids such as tryptophan halts RB division and differentiation into EBS (Beatty *et al.*, 1994). Other studies have shown that DNA replication and segregation of the cells continue but cell division is abrogated in persistent chlamydial cells (Gerard *et al.*, 2004; Belland *et al.*, 2003; Gerard *et al.*, 2001). Collectively, these results suggest that *Chlamydia trachomatis* may survive unfavourable conditions by entering a persistent state in which the cells are still viable and able to undergo DNA replication and segregation and that at least *in vitro* removal of the stressful conditions can trigger reactivation of the cell.

The final paragraph of the main body of the review discusses the implications of long-term exposure to *Chlamydia trachomatis* infections. By positioning this paragraph at the end of the main body it reminds the reader of the scope of the problem before going on to conclude (in the conclusion below) the need for further study in this area

Evidence indicates that prolonged exposure to *Chlamydia trachomatis* infections increases the risk of developing pelvic inflammatory disease (PID), ectopic pregnancy, and infertility in females, and inflammation of the prostate gland and, in some cases, infertility in males (Idahl *et al.*, 2004; Westrom, 1994; Hillis *et al.*, 1993). This prolonged exposure may be due to recurrent infection (through either re-infection or treatment failure) or through failure to diagnose the infection. The latter is likely as *Chlamydia trachomatis* infections are often asymptomatic, with up to a half of infected males and about two-thirds of infected females carrying *Chlamydia trachomatis* infections for 3 months or longer (Wilson *et al.*, 2002; Cates and Wasserheit, 1991) without being aware of being infected. In such cases, *Chlamydia trachomatis* can progress to the most internal tissues, and establish a chronic infection which can cause infertility and ectopic pregnancies.

Conclusion: summarizes the major points covered in the main body of the review and formulates questions that need further research. The conclusion maintains the focus established in the introduction, that the scope of the problem posed by *Chlamydia trachomatis* infections is increasing and therefore greater research into this area is required. Wider implications of this work are suggested (i.e. could in turn lead to more effective strategies being developed)

In conclusion, untreated and/or recurring infection poses a significant health threat, in particular to women. Furthermore, there is a growing incidence of *Chlamydia trachomatis* infections and an increase in the proportion of treatment failures. This highlights the urgent need to understand the exact mechanisms behind the organism's ability to survive antibiotic treatment. Whilst a significant body of evidence suggests that *Chlamydia trachomatis* can enter a persistent state and survive treatment, most of these studies have been conducted *in vitro*. Therefore, there is a need to conduct *in vivo* studies on the underpinning molecular mechanisms leading to persistence. Greater understanding of these mechanisms could in turn lead to the development of more effective treatment strategies.

References

All the references cited in the body of the text are listed in alphabetical order at the end of the writing in the form of a reference list and are consistent with the referencing style used in text (i.e. the author–date system). A wide range of references are used from a number of different peer-reviewed journals as well as government publications. References are cited from 1977 through to 2007, with the majority of publications taken from the last 5 years, indicating that the material covered is thorough and up to date

Beatty, W.L., Byrne, G.I., Morrison, R.P. (1994). Repeated and persistent infection with *Chlamydia* and the development of chronic inflammation and disease. *Trends Microbiol.,* **2**, 94–98.

Belland, R.J., Nelson, D.E., Virok, D., Crane, D.D., Hogan, D., Sturdevant, D., Beatty, W.L., Caldwell, H.D. (2003). Transcriptome analysis of chlamydial growth during IFN-gamma-mediated persistence and reactivation. *Proc. Natl Acad. Sci. USA*, **100**, 15971–15976.

Binet, R. and Maurelli, A.T. (2005). Fitness cost due to mutations in the 16S rRNA associated with spectinomycin resistance in *Chlamydia psittaci* 6BC. *Antimicrob. Agents Chemother*., **49**, 4455–4464.

Bragina, E.Y., Gomberg, M.A., Dmitriev, G.A. (2001). Electronmicroscopic evidence of persistent chlamydial infection following treatment. *J. Eur. Acad. Dermatol. Venereol*., **15**, 405–409.

Cates, W. and Wasserheit, J. N. (1991). Genital chlamydial infections: epidemiology and reproductive sequalae. *Am. J. Obstet. Gynaecol.*, **164**, 1771–1781.

Coles, A.M., Reynolds, D.J., Harper, A., Devitt, A., Pearce, J.H. (1993). Low-nutrient induction of abnormal chlamydial development: a novel component of chlamydial pathogenesis? *FEMS Microbiol. Lett.,* **106**, 193–200.

Dean, D., Sucherland, R.J., Stamm, S.E. (2000). Evidence for long-term cervical persistence of Chlamydia trachomatis by omp1 genotyping. *J. Infect. Dis.,* **182**, 909–916.

Department of Health. (2002). *Sexual health and HIV strategy: implementation action plan*. Department of Health, London.

Dessus-Babus, S., Bebear, C.M., Charron, A., Bebear, C., de Barbeyrac, B. (1998). Sequencing of gyrase and topoisomerase IV quinolone-resistance-determining regions of *Chlamydia trachomatis* and characterization of quinolone-resistant mutants obtained in vitro. *Antimicrob. Agents Chemother*., **42**, 2474–2481.

Dreses-Werringloer, U., Padubrin, I., Jurgens-Saathoff, B., Hudson, A.P., Zeidler, H., Kohler, K. (2000). Persistence of *Chlamydia trachomatis* is induced by ciprofloxacin and ofloxacin in vitro. *Antimicrob. Agents Chemother.,* **44**, 3288–3297.

Gerard, H.C., Krausse-Opatz, B., Wang, Z., Rudy, D., Rao, J.P., Zeidler, H., Schumacher, H.R., Whittum-Hudson, J.A., Kohler, L., Hudson, A.P. (2001). Expression of *Chlamydia trachomatis* genes encoding products required for DNA synthesis and cell division during active versus persistent infections. *Mol. Microbiol.,* **41**, 731–741.

Gerard, H.C., Whittum-Hudson, J.A., Schumacher, H.R., Hudson, A.P. (2004). Differential expression of three *Chlamydia trachomatis* hsp60-encoding genes in active vs. persistent infections. *Microb. Pathog.*, **36**, 35–39.

Health Protection Agency. (2006). General Information—*Chlamydia*. [online]. Available at: http://www.hpa.org.uk/infections/topics_az/hiv_and_sti/sti-chlamydia/general.htm#serious (last accessed 15 March 2007).

Hillis, S.D., Joesoef, R., Marchbanks, P.A., Wasserheit, J.N., Cates, W., Westrom, L. (1993). Delayed care of the pelvic inflammatory disease as a risk factor for impaired fertility. *Am. J. Obset. Gynecol.*, **168**, 1503–1509.

Idahl, A., Boman, J., Kumlin, U. (2004). Demonstration of *Chlamydia trachomatis* IgG antibodies in the male partner of the infertile couple is correlated with a reduced likelihood of achieving pregnancy. *Hum. Reprod.*, **19**, 1121–1126.

Jones, R.B., Van der Pol, B., Martin, D.H., Shepard, M.K. (1990). Partial characterization of *Chlamydia trachomatis* isolates resistant to multiple antibiotics. *J. Infect. Dis.*, **162**, 1309–1315.

Johnson, F.W. and Hobson, D. (1977). The effect of penicillin on genital strains of *Chlamydia trachomatis* in tissue culture. *J. Antimicrob. Chemother.*, **3**, 49–56.

La Montagne, D.S, Baster, K., Emmett, L. (2004). *The Chlamydia recall study: investigating the incidence and infection rates of genital chlamydial infection among 16–24 year old women attending general practice, family planning and genitourinary medicine clinics, March 2002–2004, final report, part 1.* Health Protection Agency Centre for Infections. London.

Lau, C.Y., Qureshi, A.K. (2002). Azithromycin versus doxycycline for genital chlamydial infections: a meta analysis of randomised clinical trials. *Sex. Transm. Dis.*, **29**, 497–502.

Lefevre, J.C. and Lepargneur, J.P. (1998). Comparative in vitro susceptibility of a tetracycline-resistant *Chlamydia trachomatis* strain isolated in Toulouse (France). *Sex. Transm. Dis.*, **25**, 350–352.

Misyurina, O.Y., Chipitsyna, E. V., Finashutina, Y.P. (2004). Mutations in a 23S rRNA gene of *Chlamydia trachomatis* associated with resistance to macrolides. *Antimicrob. Agents Chemother.*, **48**, 1347–1349.

Mpiga, P. and Ravaoarinora, M. (2006). *Chlamydia trachomatis* persistence: an update. *Microbiol. Res.*, **161**, 9–19.

Nanagara, R., Li, F., Beutler, A., Hudson, A., Schumacher, H.R. (1995). Alteration of *Chlamydia trachomatis* biologic behavior in synovial membranes: suppression of surface antigen production in reactive arthritis and Reiter's syndrome. *Arthrit. Rheum.*, **38**, 1410–1417.

Pantoja, L.G., Miller, R.D., Ramirez, J.A., Molestina, R.E., Summersgill, J.T. (2001). Characterization of *Chlamydia pneumoniae* persistence in HEp-2 cells treated with gamma interferon. *Infect. Immun.*, **69**, 7927–7932.

Segreti, J., Kapell, K.S., Trenholme, G.M. (1992). In vitro activity of beta-lactam drugs and sulbactum against *Chlamydia trachomatis. Diagn. Microbiol. Infect. Dis.,* **15**, 371–373.

Somani, J., Bhullar, V.B., Workowski, K.A., Farshy, C.E., Black, C.M. (2000). Multiple drug-resistant *Chlamydia trachomatis* associated with clinical treatment failure. *J. Infect. Dis.*, **181**, 1421–1427.

Suchland, R., Stamm, W., Denamur, E., Rothstein, D. (2003a). *C. trachomatis* mutants resistant to rifampicin which map in the rpoB gene did not confer significant resistance to rifalazil [abstract C1–2130]. 43rd Interscience Conference on Antimicrobial Agents and Chemotherapy (Chicago, 14–17 September 2003). Washington, DC: American Society for Microbiology.

Suchland, R.J., Geisler, W.M., Stamm, W.E. (2003b). Methodologies and cell lines used for antimicrobial susceptibility testing of Chlamydia spp. *Antimicrob. Agents Chemother.,* **47**, 636–642.

Westrom, L.V. (1994). Sexually transmitted diseases and infertility. *Sex. Transm. Dis.,* **21**, 532–537.

Whittington, W.L.H., Kent, C., Kissinger, P., Oh, M.K., Fortenberry, J.D., Hillis, S.E., Litchfield, B., Bolan, G.A., St Louis, M.E., Farley, T.A., Handsfield, H.H. (2001). Determinants of persistent and recurrent *Chlamydia trachomatis* infection in young women. *Sex. Transm. Dis.,* **28**, 117–123.

Wilson, J.S., Honey, E., Paavonen, J., Mardh, P.A., Stary, A., Stray-Pedersen, B. (2002). A systematic review of the prevalence of *Chlamydia trachomatis* among European women. *Hum. Reprod. Update,* **8**(4), 385–394.

References

Aveyard, H. (2007). *Doing a literature review in health and social care: a practical guide.* Open University Press, Maidenhead.

Bourner, T. (1996). The research process: four steps to success. In: Greenfield, T. ed. *Research methods: guidance for postgraduates.* Arnold, London.

Bruce, C. (1994). Supervising literature reviews. In: Zuber-Skerritt, O. and Ryan, Y. eds. *Quality in postgraduate education.* Kogan Page, London.

Cooper, H.M. (1982). Scientific guidelines for conducting integrative literature reviews. *Rev. Educ. Res.,* **52**(2), 291–302.

Hart, C. (2005). *Doing a literature review: releasing the social science research imagination.* Sage, London.

Chapter 8

Writing a research proposal

Introduction

The research proposal (also termed the grant proposal) puts forward a clear research question or hypothesis and then sets out an experimental strategy for investigating the hypothesis or question within a defined period of time. You will almost certainly be expected to write a proposal at the beginning of your Master's or PhD research programme, and it is a skill that you will continue to develop and use as a research scientist. The aim of this chapter is to guide you through the process of writing a well-planned and well-structured research proposal.

Specifically, this chapter will:

- summarize the typical content of each component of the research proposal
- describe a strategy for writing the proposal, starting from defining the scope of your project through to submitting the completed proposal
- describe how you can find funding sources for bioscience research
- describe how applications submitted to funding organizations are peer reviewed and the decision for allocating funding reached
- present an annotated example of a research proposal.

8.1 Components of the research proposal

If you are writing a research proposal as a Master's or PhD student, then the objective of the proposal is to outline the work you will undertake during the course of your research project. The advantage of writing a proposal before you start conducting your experimental work is that you will become acquainted with the background to the research area and with the methods that are commonly used within the subject. This in turn will help you establish a context for your work and assist in planning the experiments you will do during the project. Overall, you are more likely to produce good work and less likely to waste time and resources.

The research proposal written as part of a Master's or PhD programme of study is similar in content and format to the research proposal (commonly termed a grant proposal) which is submitted to a funding organization to obtain funds to support a programme of research. The main difference is that the Master's or PhD research proposal is likely to be shorter and will not include supporting documentation, such as the curriculum vitae (CV) of researchers and letters from collaborators, which normally forms part of the proposal submitted to a funding organization.

Writing grant proposals is an essential part of a research scientist's role as you cannot pursue your research unless funds are secured to finance the experimental work. Proposals are generally submitted by groups of two or three investigators. However, it is now becoming increasingly common to submit proposals as teams of researchers, collaborating on projects from different disciplines and even different countries. Hence compiling a strong application requires extensive planning and communication between the researchers. In addition, where the work will be relevant to users outside the scientific community, such as the health service, then discussion with the potential end users will also be necessary. Funding organizations receive a large number of applications and therefore securing funding is a highly competitive activity. If your proposal sets out clear research aims which are well justified in the context of the background literature, outlines a well-designed experimental strategy, and is written in clear and concise language, it will stand a better chance of being selected to receive funding. The training you receive in writing a research proposal for your Master's or PhD project will begin to develop your ability to produce well-planned and well-written research proposals.

A complete research proposal submitted to a funding organization typically consists of:

- A description of the proposed research project which includes the hypotheses and aims of the project, the background to the work proposed and its significance, a detailed experimental work plan which explains how and when the aims will be accomplished, and an explanation of how the results will be disseminated and shared.
- A description of how the ethical requirements of the work (if any) will be fulfilled, with supporting documentation where necessary.
- A budget which sets out the costs for each year of the project and justifies the resources requested.
- An outline of the research experience of the applicants and any key workers who will contribute to the work.
- An outline of the research facilities available at the host institution (that is, the place where the research will be conducted) to support the project.

The typical content of a research proposal is outlined below. Remember, you may need to adapt the content described here to fit the specific guidelines provided by your programme of study. Similarly, when applying to a funding organization, you will need to adapt the content to fit with the specific instructions supplied by the funding scheme. For detailed examples of instructions supplied by funding organizations to prospective applicants, consult the following sources of information: Medical Research Council (2007), European Research Council (2008), National Science Foundation (2008).

8.1.1 Title of the proposed work

The title should concisely and accurately describe the essential content of the proposed work. When writing the title, include keywords that are strongly associated with the content of the article and avoid using abbreviations and non-essential terms such as 'a study to investigate . . .' When writing your title, always keep within the word limit stipulated in the instructions supplied. For an example of a concise and is informative title, see the sample research proposal included in section 8.5.

8.1.2 Abstract

The abstract is a concise summary which conveys clearly to the reader the key messages of your proposal. The abstract that forms part of a funding proposal typically contains three types of information: it outlines the research question, the main methods that will used to investigate the research question, and the significance of the proposed work. Detailed guidance on writing abstracts is provided in Chapter 10 and is summarized briefly here:

- Include the essential content expected in an abstract—that is, the research question, the methods proposed, and the significance of the work. Arrange this information logically so that it is easy to follow.
- Make the abstract comprehensive so that it can be understood without reference to the main proposal.
- Avoid using abbreviations and references, and do not include illustrations.
- Keep within the word limit stipulated.

For an example of an abstract, see the sample research proposal included in section 8.5.

8.1.3 Background and significance of the proposed work

The background clearly sets out your research question, critically reviews the background literature to the research area, and explains the significance of the work proposed. Overall, the background should set your work in context and convince the reader of the need to investigate the question you are proposing. Chapter 7 describes how to write a critical review of the scientific literature which you should refer to as guidance for writing the background and significance of your proposed work.

Briefly, when writing the background to your work:

- Describe the current state of knowledge in the field. This should include a review of the recent and current developments that specifically relate to the topic under investigation. You should clearly establish what is known and what is unknown about the topic area.

- Describe the specific gap in knowledge you propose to address and why it is important and timely to study. Here you should describe the problem you intend to study and how it fits in with the existing literature. Then explain what the significance of the findings could be. For example, will it advance our conceptual understanding of a particular model or will it have practical applications such as improving diagnostic or therapeutic methods or improving agricultural practices? When describing the significance of the findings, consider who the work will benefit, how it will benefit them, and why it is important to study the problem now.
- Commonly, proposals are based on existing work for which preliminary data are available. Therefore, preliminary results should be included either in the background to the work proposed or at an appropriate place in the experimental strategy (section 8.1.5).
- Conclude with a clear statement that describes the overall aim (or hypotheses) of your proposed research.

For an example of a background and significance to a proposed research programme, see the sample research proposal included in section 8.5.

8.1.4 Objectives of the proposed work

Objectives are specific statements which concisely describe the outcomes you expect to achieve. Each objective should address a specific part of the research question and collectively should answer the research question the project is proposing. To understand this, consider your research question as a broad statement of what you expect to achieve (also termed the aim of the project). You can then break down the research question into smaller, more specific statements (or sub-questions) which collectively describe what you will do to achieve your research questions. When writing the objectives:

- Formulate each objective so that it can be tested **experimentally** and has a **measurable** outcome.
- Describe each objective succinctly in one or two sentences only.
- Be realistic about what can be achieved during the time span of the project. A programme of research lasting 3 years should typically consist of 3–5 objectives so that the work can be completed within the time scale of the project.

For an example of how the objectives of a proposed research programme can be written, see the sample research proposal included in section 8.5.

8.1.5 Experimental strategy

This section describes **which methods** you will use to accomplish each objective, **why** you will use these methods, and **when** you will start and complete each objective. This section will be used by the reader to judge if the proposed experimental strategy is robust and will answer the research question posed. Therefore, when writing the experimental strategy, include sufficient detail to allow the reader to make a judgement about the quality of your experimental design.

When writing the experimental strategy:

- Include a detailed description of the experimental methods and provide a rationale for why you are using them. You can structure this section by writing each objective as a sub-heading and then describe underneath the sub-heading the experiment or set of experiments you will conduct to achieve the objective.
- Provide a clear rationale for the methods you propose to use and include precise details of the experimental procedures. Include information such as the sample size, the controls, how the samples will be collected, any treatments you will apply, how the data will be collected, and how the data will be analysed. For each experiment you describe, include brief details of the results you expect to obtain. If your data will be analysed statistically, then describe the statistical methods you intend to use.
- The amount of detail you include for each experiment will depend on whether the method you are describing is a standard protocol or a new method. If it is a standard method, such as SDS–PAGE or PCR, do not provide detailed descriptions. However, if your methodology is new, then you should provide more detailed descriptions. Include a flow diagram showing the sequence of experiments if it will help the reader to understand your experimental strategy better.
- Include a timeline which shows when you will start and complete each objective. To do this, first arrange your experiments in the order you will tackle them. Group together the experiments you will work on simultaneously and place sequentially those experiments which can start only after the preceding ones are complete. Follow this by estimating the time it will take to complete each experiment. You could present the timeline graphically in the form of a Gantt chart or a flow chart. An example of a Gantt chart showing the timeline of a research project is included as part of the annotated example of a research proposal in section 8.5. A timeline of when you expect to complete particular parts of your research and what you will achieve by that time is an important aspect of the plan as it provides milestones for measuring how your research is progressing.
- Include a description of potential problems that could arise during your research programme and provide alternative approaches that could be used to address each problem should it occur. This is particularly important for experiments that rely on data generated from an earlier experiment and therefore cannot be undertaken unless the preceding experiment is successful. Outlining alternative approaches is an important part of your plan as it demonstrates that you will be able to continue with the project should any parts of the experimental strategy fail.

For an example of how the experimental strategy section of a proposal can be written, see the sample research proposal included in section 8.5.

8.1.6 Resources required and justification for the requested resources

This section should include a budget listing all the items required to complete the project and the costs associated with each item. In addition, you will be expected to justify why you are requesting these items. Most funding bodies provide detailed guidance on preparing budgets and you should follow the guidelines provided exactly. Some general points to assist you in preparing budgets are:

- First, go through your research proposal and make a list of the people who will do the work, and the equipment and consumables you will require for each experiment. Include the amount of time existing staff will contribute weekly to the project in activities such as supervising the project and writing reports, and new staff that you will employ such as research fellows or technical staff. Include the cost of running existing equipment as well as purchasing new equipment. Other items could include consumables, travel and subsistence for attending conferences, travel to field sites, cost of licences and permits, and publishing costs.
- Find out the cost of each item on your list. You should be able to seek advice from your institution on costs of staff salaries and from relevant suppliers for consumables and equipment.
- When presenting your budget, show the information by category (for example, consumables, travel, equipment, salary) for every year of the project. Compile the budget in a spreadsheet to help you calculate the sub-totals and totals of expenditure easily.
- Include a justification for the resources you are requesting. For example, if you are requesting funds to employ a research assistant, then describe the work the new member of staff will conduct. If you are requesting travel costs, explain why you will be travelling and where you will be travelling to.

When writing the budget, make sure all the items are listed and the costs and calculations are accurate. Be realistic about the amount of funding you are requesting. You do not want to under-budget and then find that you run out of money and cannot continue with your project. Nor should you over-budget in a way that is not an honest representation of how much the project will cost.

For an example of a budget and justification for the resources requested, see the sample research proposal included in section 8.5.

8.1.7 Ethical considerations

If your work involves human subjects and tissues, animals, genetically modified organisms, dangerous pathogens, controlled drugs, or protected wildlife and habitats, then you will require ethical approval or a licence (or both) to conduct the work. If you are entering directly into a funded project, then regulatory and ethical approval will have been dealt with at the funding stage. However, you will still need to be aware of, and comply with, the regulations and procedures relating to the ethical and health and safety aspects of your work. If you are applying for funding, then the approval or licence may be necessary before an application is considered by a funding body, or it may be possible to obtain after submitting the grant proposal but before the award is confirmed. In the latter case, you will need to describe the arrangements in place for obtaining the approval and licences should you be awarded funding. You will be able to get guidance on regulations governing work with any of the materials listed here and on procedures for obtaining approval or licences from your institution. Additional sources of information are:

- Regulations relating to work with wildlife and habitats are set out in the UK in a number of Acts such as the Wildlife and Countryside Act 1981 (and subsequent amendments) and the Countryside and Rights of Way Act 2000. Information can be accessed from the Joint Nature Conservation Committee website (2008).
- Regulations relating to work with human subjects (including human material and records) are embodied in the Declaration of Helsinki (World Medical Association, 2004) which was first adopted in 1964 and has subsequently been revised five times, most recently in the year 2000.
- Regulations relating to work with human tissue in the UK are set out in the Human Tissue Act 2004 (Department of Health, 2007a).
- Regulations relating to work with animals (in the UK) are set out in a number of Acts of Parliament, principally the Animals (Scientific Procedures) Act 1986 and associated Codes of Practice relating to care and housing of laboratory animals, including breeding and humane killing. This information is available from the Home Office (2008a).
- Regulations relating to the use of genetically modified organisms are set out in the UK in The Genetically Modified Organism (contained use) Regulations 2000 (and subsequent amendments) and are available from the Health and Safety Executive (2008).
- Safeguards recommended when working with dangerous pathogens in the UK are set out by the Advisory Committee on Dangerous Pathogens (1995 and subsequent recommendations). The information is available from the Department of Health (2007b).
- Regulations relating to the use of controlled drugs in the UK are set out in the Misuse of Drugs Act 1971 (and subsequent amendments). This information is available from the Home Office (2008b).

8.1.8 Details of the applicants and the environment in which the research will be conducted

An important criterion on which the decision to allocate funding is made is whether the applicant is able to complete the work successfully based on their research expertise. This section should therefore clearly set out why you and your co-investigators are the best qualified to undertake the research. This information is usually supplied in the form of a CV and includes a list of your publications, research funding currently held or previously secured, and any other information which shows you have a strong career record. Examples of additional types of information could include invited presentations to conferences or the organization of international conferences.

A further consideration when allocating funding is whether the applicant has access to resources which will ensure that there is a high likelihood of the project being completed successfully. You should therefore clearly set out the details and skills of key research staff who will be contributing to the project, details of any collaborators and the specific expertise or materials they will provide, and a description of the main and relevant research facilities that are available to use at your institution.

8.1.9 Dissemination of research results and data sharing

Your research proposal should demonstrate that you are aware of the potential value of your results (that is, who the work will benefit and how it will benefit them) and that you are willing to share the information with other interested parties. Your proposal should therefore describe how the results arising from your research will be disseminated amongst the scientific community and other parties who may benefit from your findings. This could include publication in peer-reviewed journals, but also how the information will be communicated to other interested groups outside the academic community such as policy-makers, teachers, industry, and the health services.

Your proposal should also contain a brief statement outlining your plans to share the data associated with your findings with the wider scientific community. Chapter 3 (section 3.3) describes the different ways in which scientific data are shared and the purposes of sharing the data, to which you should refer. In summary, as a research scientist you have a responsibility to report your results and share the associated data sets with the wider scientific community in a timely manner. This exchange of information facilitates scientific discovery and is therefore of benefit to both the researchers and the public. There are a number of ways in which data can be shared and this will depend on the type of data your study generates. You could publish your results in open access peer-reviewed journals or deposit your e-prints in a repository or the material generated as part of your study into publicly accessible databases (see Chapter 3, section 3.3). If your data, or some aspects of them, cannot be shared for some reason, then you should explain why. For example, if your work involves human subjects or your findings could be commercially exploited, then you should explain this as the reason for not sharing data.

8.1.10 References

Your reference list will include all the references cited in your proposal. You should cite references that are **current** and **relevant** to your work. This will show the reader that you are fully aware of how the topic area is developing. Research proposals commonly include unpublished preliminary observations which the researcher proposes to develop further. If your proposal includes unpublished data then cite the source appropriately in your work (see Chapter 1, Table 1.1).

8.2 Writing the proposal

A well-written research proposal requires careful planning. If you plan well, then your written proposal is more likely to present a logical and well-thought-out case for supporting your work. The three stages of writing a research proposal—planning, writing, and revising—are described below. Before starting to plan, you must familiarize yourself with the guidelines supplied by your programme of study (if you are writing a proposal as part of your Master's or PhD programme) or the funding organization (if you are writing a proposal to submit to a funding body). The guidelines will be detailed and include information on:

- The components and content of the proposal.
- Any supporting documentation that should be supplied at the same time as the proposal such as CVs for each investigator, estimates of equipment costs, and supporting letters from collaborators.
- Ethical requirements if your work involves human subjects and tissues, animals, human fetal material, genetically modified organisms, dangerous pathogens, controlled drugs, or protected wildlife or habitats.
- The page limits of the proposal and whether appendices are allowed or not.
- Word limits for the title and abstract.
- The formatting you should use, including font style, size, and margins.
- How the proposal should be submitted.

You should comply with the instructions supplied **exactly.** If you do not, and your proposal is being assessed, you will lose marks unnecessarily. If you are submitting an application to a funding body then your application may be unsuccessful or rejected outright. It is a good idea to read through examples of well-written research proposals to familiarize yourself with the content, structure, and style expected before you start to write. An annotated example of a research proposal is included in section 8.5 and you should also be able to obtain further examples from your supervisor or colleagues.

8.2.1 Stage 1: plan your research proposal

The first step in writing a research proposal is to plan an outline of your project to cover the various sections outlined in section 8.1. Break your plan into two parts. In the first part, define clearly the specific problem you propose to investigate and the background to it. In the second part, plan the experimental strategy you will use to answer your question. Before you begin to plan, conduct an exhaustive literature search in the topic area related to your investigation (Chapter 5) to familiarize yourself with the background and methods relating to the topic. This will assist you in formulating your research question and your experimental strategy. To plan an outline of your proposal, work through the following questions (adapted from National Institutes of Health, 2008):

Part 1: define your research question and position it within the background to the topic area

1. **Which specific question do you propose to address?** Write down your research question concisely. Alternatively, you can state your research problem as a hypothesis.
2. **Which gap in knowledge does your research question address?** Write a brief paragraph summarizing what is known and what is unknown in the field and identify the specific gap in knowledge your research question will address. This will help you to position your work clearly in the context of the existing literature.
3. **Why is it important to study the question?** Again, write a brief paragraph describing why there is a need to study this question and the impact your work will have. Consider who your work will benefit and how it will benefit them.
4. **Why is it important to study the question now?** Write a few sentences describing why it is important that the study should be conducted now. Consider how the work fits into current local, national, or international agendas.

Your research question should be original, and add to and extend the existing body of knowledge in a significant way. It should also be timely to study. By the end of this planning section, you should be able to see to what extent this is true for your proposed work. If your study does not fit these criteria, then you will need to revise your proposal.

Part 2: design your experimental strategy

1. **What are the objectives of the study?** Break down your research question into 3–5 objectives (or sub-questions). Write each objective in one or two concise sentences and formulate each objective so that it is testable experimentally and has a measurable outcome. Check that your objectives collectively answer your research question.
2. **Which experimental methods will you use to achieve the objectives?** Write each objective as a sub-heading and underneath the sub-heading write the experimental methods you will use to accomplish the objective. For each experiment write down the result you expect to achieve. Review your experimental methods and check that (1) the method is appropriate for achieving the objective, (2) there

is a clear relationship between the objective, methods, and expected outcome, and (3) the experiments are arranged in a logical order.

3. **How long will it take to accomplish the objectives?** Go through your methods and estimate the time it will take to complete each set of experiments. Based on these estimates, produce a timeline showing the objectives, the sequence of experiments, and the start and end times for each objective.
4. **What resources do you require to complete the work successfully?** Again, go through your methods and make a list of the key people who will do the work and a list of the equipment and consumables you will require for each experiment. Some of the resources you may already have, others you will need to request funding for. Specify the approximate cost of each item you require funding for.
5. **Does your work require a licence or ethical approval?** Identify the ethical requirements of your proposed work. Take advice from your institution if you are unaware of the procedures and regulations. This will help you plan in advance whether you need to apply for any licences or ethical approval before you can start work . If you are entering directly into a funded project, then regulatory and ethical approval will have been dealt with at the funding stage. However, you will still need to be familiar with the regulations and procedures relating to the ethical and health and safety aspects of your work and comply with them.

This initial outline should be no more than two pages in length but at the end you will have a focused research question and the beginnings of an experimental strategy. If you are applying for funding, you will have a clearer idea of the amount of funding you are looking for and the length of time it will take to complete the research successfully. This in turn will help you identify the right funding source for your project (section 8.3).

8.2.2 Stage 2: write the proposal

As you write, you will flesh out the outline you produced during the planning stage. Use the typical content and guidelines provided in section 8.1 to guide you in your writing. However, remember, that the guidelines for each programme or funding scheme are different and you should therefore adapt the content to match the specific guidelines you have been provided with. It is also a good idea to keep in mind the criteria by which your proposal will be judged (section 8.4). Overall, the reviewers will be looking for focused research questions (or hypotheses) which are justified in the context of the background literature and which will make a significant contribution to the field. They will be looking for a well-designed and well-justified experimental strategy which shows a clear link between the aims of your research, the methods proposed, and the outcomes expected. They will also be looking to see that you can complete your work within the time scale and with the resources you have outlined.

To achieve this within the page or word limit of the proposal, you will need to select carefully the information you include in your proposal. Choose only that information which is necessary to understand your case. Write concisely and do not be tempted to reduce the font size below that stipulated in your guidelines.

A suggested order for writing the proposal is:

1. Write the research question (or hypotheses) and objectives of the project.
2. Write the experimental strategy.
3. Write the background and significance of the work.
4. Finish with the title and abstract.

Start compiling applications for obtaining ethical approval and licences (where necessary) as soon as the experimental strategy section is complete. Licences and approvals take time to obtain and you do not want to delay your application or your work unnecessarily. You can also start compiling the budget and any outstanding sections and supporting documentation once the experimental strategy is written.

When writing your proposal:

- Talk to your supervisor, co-applicants, or colleagues about your research plans. This will assist you in developing your ideas and your time scales.
- Conduct up-to-date literature searches to make sure that the information you are including is the most current.
- Write concisely and clearly. Use short sentences and avoid needless repetition. Try to use simple words as far as possible and avoid unnecessary technical jargon. Sometimes technical words are necessary, of course, but on the whole avoid complex terms. Where abbreviations are used, define them at the point of first use.
- Use the future tense when writing a proposal. For example, say '*we will carry out . . .*' instead of '*we carried out . . .*'.
- Use sub-headings to break up sections of your application so that it is easier to follow.
- Include illustrations if it will help the reviewers understand your work better. This could include timelines in the form of Gantt charts or flow charts to show the sequence of experiments, or figures or tables showing any preliminary data, or line drawings to illustrate concepts or models.
- When writing a budget, be accurate and honest about the costs. The budget is an important part of your proposal and the reviewers will look through it carefully to see whether the proposed research is 'value for money'.
- If writing an application to submit to a funding organization, explain how the proposed work matches the funding priorities of the organization. Avoid statements such as 'this study will eventually lead to a cure' (if, for example, the scheme is funding health-related research). Instead, describe what the short-term and long-term outcomes of the project will be and how these fit into and extend our current understanding of the problem.

8.2.3 Stage 3: review drafts of your proposal

Once the first draft is written, review it for factual accuracy, completeness of content, organization and coherence. Use the checklist provided in Box 8.1 to assist you in reviewing the draft. It is also a good idea to review your draft against the assessment criteria the reviewers will use to judge the proposal (section 8.4) to see how it measures up. This will help you identify areas that require refining or adding to. It is always useful to ask a colleague familiar with the subject matter to look through and provide constructive comments on the proposal.

BOX 8.1 Reviewing a research proposal—checklist

If you are writing a research proposal which is part of your Master's or PhD programme of study and not for submission to a funding organization, then omit the parts that are not relevant to you such as letters from collaborators and CVs.

Title and abstract

- ❑ Does the title conform to the length stipulated by the instructions supplied?
- ❑ Is the title descriptive of the content of the proposal?
- ❑ Does the abstract conform to the length limit stipulated in the instructions supplied?
- ❑ Does the abstract outline your research question, methods, and significance of the study arranged in a logical order?
- ❑ Are there any unnecessary words or sentences in the title or abstract that can be removed to improve conciseness and clarity?

Background and significance of the proposal

- ❑ Do you have a clear hypothesis or research question?
- ❑ Does your background set out the most relevant information so that the reader can place your work in context?
- ❑ Does your background show clearly which gap in knowledge you are addressing?
- ❑ Does your background show clearly why it is important to study the research question and why it should be studied now?
- ❑ Is your research question aligned to the funding priorities of the funding organization?

Objectives of the proposed study

- ❑ Do you have clear objectives that address the research question or hypothesis?
- ❑ Can each objective be tested experimentally and does it have a measurable outcome?

Experimental strategy

- ❑ Have you selected the most appropriate methods for testing your research questions?
- ❑ Does your description of methods contain sufficient details for the reviewer to judge that your proposed strategy is robust?
- ❑ Have you described how you will analyse your data?
- ❑ Have you described what outcome you expect for each experiment?
- ❑ Is there a clear relationship between the aims of the project, the methods, and your proposed outcomes?
- ❑ Can your objectives be realistically achieved within the time scale you have outlined?
- ❑ Have you identified any potential problems and provided suggestions for overcoming the problems?

Ethical considerations

- ❑ Is your proposed work ethical?
- ❑ Have you obtained ethical approval and licences (or included details of arrangements in place to obtain approval and licences) where necessary?

Budget and justification of resources

- ❑ Is your projected budget accurate?
- ❑ Have you justified the resources you have requested adequately?
- ❑ Can you realistically complete the work with the resources requested?
- ❑ Does the budget match the amount of funds the organization will pay?

Details of staff and environment

- ❑ Have you included details of your research experience that show that you are competent to complete the project successfully?
- ❑ Have you included details about your research environment and the skills of the key workers who will contribute to the project to show that the likelihood of completing the project successfully is high?
- ❑ If you and your institution do not have a particular material or expertise, have you provided supporting letters from collaborators who will supply the material or expertise?

Dissemination and data sharing

- ❑ Have you included a statement which describes how you will disseminate your findings to the academic community and other interested parties outside the academic community?
- ❑ Have you included a statement that describes how you will share the data generated through your research with the scientific community?

BOX 8.1 Cont'd

References

- ❑ Are your references relevant to the proposed work and are they up to date?
- ❑ Are all the sources you have used appropriately acknowledged in the text?
- ❑ Are all the sources cited in the text also listed at the end in the reference list?
- ❑ Have you used a consistent style of referencing both in the text and in compiling the reference list?
- ❑ Does the format of referencing used conform to any guidelines provided?

Organization and coherence of your proposal

- ❑ Have you used appropriate sub-headings to section the proposal?
- ❑ Does each paragraph cover one topic only?
- ❑ Do the paragraphs follow in a logical manner so that the line of reasoning is clear?
- ❑ Are appropriate transitions used to link each successive paragraph?
- ❑ Are there any unclear or long sentences that can be rewritten to improve clarity?
- ❑ Are there any places where you have repeated what you have said elsewhere and therefore where content needs deleting?
- ❑ If you have used illustrations, are they cited in the text?
- ❑ Is each illustration numbered sequentially?
- ❑ Does each illustration include a title, legend, and a number?
- ❑ If you have used any abbreviations, symbols, or units, are they in the correct format and consistent in style?
- ❑ Have you defined abbreviations at first use?
- ❑ Have you checked for mistakes in spelling, grammar, and punctuation?
- ❑ Have you checked the layout of the document (e.g. font size and style, spacing, margin size) and the length to see that it conforms to the guidelines provided?

Once you have reviewed and redrafted the proposal, compiled any supporting documents such as CVs, and obtained the appropriate signatures then you are ready to submit your application. Make sure you submit it by the deadline and in the correct format.

8.3 Funding sources for bioscience research

Although you will not be applying for funding until a later stage of your career, it is worth having an overview of who funds bioscience research and how funding is allocated (section 8.4).

Research funding for the biosciences can be obtained from a variety of sources. These include research councils, charitable trusts, industry, and government departments. If you are looking for funding sources, seek advice from colleagues and from the research office at your institution. In addition, you can search through the following websites:

- **Research Fortnight Online** (http://www.researchresearch.com/). This is a web-based information service on funding opportunities and research news. It contains a searchable database of worldwide funding sources.
- **GrantsNet** (http://www.grantsnet.org/). This website, hosted by the journal *Science*, lists funding opportunities from Europe, Asia, and the Americas.
- **European Community Research and Development Information Service (CORDIS)** (http://europa.eu/index_en.htm): CORDIS provides information on European Union Research funding and research activities.

Most countries will also have country-specific websites which list country-specific funding sources. For example, the UK has **Research Councils UK (RCUK)** (http://www.rcuk.ac.uk/default.htm). This website acts as a portal to the seven council websites which fund research in the UK. Australia has the **Australian competitive grants register** (http://www.dest.gov.au/sectors/research_sector/) which links through to research funding sources in Australia. You will be able to find out about similar country-specific websites by consulting colleagues.

When applying for funding, read the information supplied by the funding organization carefully to see:

- What areas of research are funded.
- The types of award available.
- The career level required for the different award types.
- The maximum levels of funding available for the different awards.

This will help you identify the right funding source for your research project. Remember, your application is more likely to be successful if your research objectives and your level of research experience align closely with the funding priorities of the organization and the award type.

8.4 Peer review and outcome of your application

A grant proposal submitted to a research funding organization will pass through a three-stage review process. On receipt of an application, the proposal undergoes an initial evaluation by a subject specialist who is a member of the funding body. If the proposal is considered weak or does not fall within the research funding priorities of the organization, then it will be rejected at this stage. If the proposal passes the initial review, it will be sent out for external peer review. External reviewers will assess the quality of the application and grade it against a set of standard review criteria supplied by the funding body.

Reviewers are typically asked to comment on:

- The importance and originality of the work and how it fits in with the funding priorities of the organization.
- The strengths and weaknesses of the experimental design.
- The ability of the applicant realistically to complete the work within the time scale proposed and with the resources requested.
- The ability of the applicants to complete the work successfully based on their research expertise, collaborative links, and access to research facilities at the institutions where the work will be conducted.

Detailed review criteria against which proposals are assessed can usually be accessed through the funding organization website. For examples of review criteria consult the Natural Environment Research Council (2008) and the National Health and Medical Research Council (2008).

If strong support is not received from the reviewers, then the application is rejected. If strong support is received, then the application is sent to a grant review panel for a third and final review. Before the proposal is considered by a review panel, the feedback and recommendations received from the reviewers are passed to the applicant. The applicant may accept all of the suggestions made for improvement and adjust the proposal accordingly or, if they do not agree with aspects of the referees' comments, counter-respond with reasoned arguments stating why the suggestions have not been implemented.

Following this, the review panel (consisting of experts in particular subject areas) meet to consider the reviewers' appraisals and the applicant's response, and to rank the applications set in front of them. A final decision is then made on which proposals will be funded and which rejected.

As in the peer-review system described in Chapter 4 (section 4.2), reviewers must provide an expert review and honest, objective, and timely feedback if the system is to work effectively.

- **Expert review**. Reviewers are selected on the basis of their detailed knowledge of the subject area. This is essential as the reviewer is expected to pick up work that is not important enough, poorly designed experiments, and over-ambitious research plans which cannot be completed within the time scale of the project.
- **Confidentiality**. Reviewers are expected to maintain confidentiality about the content of the proposal they are reviewing and not discuss the work with other people unless permission from the funding body is granted; for example, when a colleague is assisting in the review process.
- **Objectivity and fairness**. Reviewers are expected to be objective and fair in their appraisal of the work. This means not allowing any sources of bias—personal, financial, or professional—to influence the evaluation. An example of when bias may creep in is if the work under review relates directly to that of the reviewer. In this case the

reviewer may be tempted to block the proposal by providing a discouraging review of the work, or to plagiarize the ideas contained within the proposal. In such cases, the scientific community expects that the reviewer will not undermine the integrity of the peer-review process by acting dishonestly.

- **Timely return of the feedback**. Reviewers are typically given a time span within which the review should be completed and the report returned to the funding body. Timely feedback is important so that the process of allocating funds runs smoothly.

8.5 Annotated example of a research proposal

The sample proposal presented in this section is a description of a 3-year research proposal and includes:

- title
- abstract
- background to the work proposed and its significance
- detailed research plan which explains how and when the work will be accomplished
- budget and justification for the resources requested
- reference list.

This proposal was submitted to a research funding organization in 2002 and is included here with permission from the author (Grahame, 2002). The annotations describe the key content and structural features of the proposal. Do not worry if you are unfamiliar with the subject matter of this proposal. By reading through the example and the annotations you should be able to apply the underlying principles described here to write and organize your own research proposal. Remember, you should always write your proposal to align with the specific guidelines you have been provided with.

DNA polymorphisms and postglacial recolonization of European shores

> The title is brief and concise, describing the topic of the proposal

Abstract

Our objectives are to characterize the DNA polymorphisms in a suitable length of mitochondrial DNA from 5 species, across a geographical range from northern Norway to the Iberian peninsula (including 25 locations). We will use this information to make inferences about the pattern of recolonization of northern European shores after the last ice age, with particular reference to the British coast. This will represent the first study of the shoreline complementary to the body of work which now exists for the land surface.

> The abstract concisely summarizes the key messages of the proposal: outlines the research question, the main methods that will be used to investigate the research question, and the significance of the proposed work

It will show how populations responded to the climatic perturbation at the end of the last ice age. In particular it will allow tests of hypotheses concerning the distribution of refugia for different species. We believe this work is timely given the current level of interest in the effects of climate change.

The proposal starts by establishing the background and significance of the pro–posed work

Background and significance of the work

The general framework

The general framework Introduces the background to the study and outlines the current understanding of the subject area

At the height of the last glacial episode, ~20 000 BP, large ice caps covered many regions of northern Europe. Regardless of how much of what is now ice-free land was then ice-bound—and there is debate over this—there is no doubt that this glaciation, like others before it, marked a major environmental counterpoint to the present. With climatic warming as the glaciation waned, species which had retreated to the southern peninsulas of Europe expanded northwards towards what are their current distributions (Hewitt, 1999). Much can be inferred about the likely patterns of recolonization involved in this expansion. Since a degree of genetic divergence occurred in each of the major southern refugia, descendant populations can be discriminated and identified as to their likely origin. Three major patterns have been inferred among the organisms which have so far been studied from this point of view.

In the first, northern populations are all held to be descended from those which survived the glaciation in a Balkan refugium. In the best-known example, the grasshopper *Chorthippus parallelus*, expansion is argued to have been rapid—this is deduced from sequence similarity and low haplotype diversity in northern populations across a broad range (Hewitt, 1999). Expansion from the Iberian and Italian peninsulas is likely to have been inhibited by the presence of mountain barriers and by the rapid arrival of the colonizers from the Balkans. The modern peninsular populations hybridize with their neighbours from the northern populations, forming hybrid zones which have received considerable attention (Hewitt, 1988; Butlin *et al.*, 1991), partly because they allow the study of populations which have evidently begun the process of speciation (Butlin, 1998).

In contrast, in hedgehogs (*Erinaceus* spp.) there were probably three major sources of northern European populations, one each in the Iberian and Italian peninsulas, with a third in the Balkans. Again, in the case of the brown bear (*Ursus arctos*) northern Europe was recolonized from Iberian and Balkan refugia, with the Alps proving a sufficient barrier to block expansion northwards from Italy until the habitat had already been filled with populations derived from both eastern and western expansion routes. Each of the species mentioned is representative of a larger group of organisms behaving in much the same way, so that Hewitt (1999) writes of the 'grasshopper, hedgehog and bear paradigms'.

Other authors discuss the possibility that in addition to the existence of refugia in the southern peninsulas, there may have been northern refugia from which at least some current distributions draw some of their populations (Stewart and Lister, 2001). These 'cryptic refugia' may have been smaller than the major ones further south, and there is controversy over whether the patterns suggesting them might in fact be due to human transport.

Discusses competing views that exist in the subject area

In contrast to our state of knowledge about recolonization of the land surface, much less attention has been paid to the characteristic assemblages of organisms in the intertidal of rocky shores. This is paradoxical given the relative simplicity of the habitat, which is linear, in contrast to the two-dimensional nature of land surface habitats. One species which has received attention is the predatory intertidal gastropod *Nucella lapillus*, the dog-whelk. Here shell shape is known to vary considerably in ways considered adaptive for resisting crab predation, and there are several instances in Britain of populations being the 'wrong' shape for the local conditions. It has been suggested that the biogeography of shape, in particular the distribution of these anomalies, may be explained in terms of re-expansion of populations isolated into discrete areas in the last ice age (Cambridge and Kitching, 1982; Crothers, 1985b; Crothers, 1992).

Identifies gaps in the knowledge base as an opportunity to conduct research

Ice age conditions have undoubtedly had a profound effect on north Atlantic intertidal biota (Wares and Cunningham, 2001). Of the six species studied in detail by these authors, five can be shown to have recolonized North America from Europe after the last ice age. Of these five, the barnacle *Semibalanus balanoides* did survive in North America, but current populations there contain a Europe-derived clade as well. While ice itself will have had a profound effect on shorelines, there is another factor to be taken into account for intertidal species, namely change of sea level. The 'inundation map' for the European shelf around Britain shows that at the height of the last glaciation, much of what is now the North Sea and all of the English Channel were dry (Pirazzoli, 1996).

Given that much of northern Britain was occupied by ice, and the English Channel and Irish Sea did not exist, this suggests that a major reorganization of the British intertidal has indeed occurred since the last ice age. The most extreme interpretation might be that many species would have been driven to extinction on the modified coastline either by cold or by ice-scour. Yet, the possibility that ice-free islands persisted through the last glacial maximum as far north as northern Norway (Dawson, 1992) means that for the shore as for the land, we cannot as yet be certain about the distribution of likely refugia. Transatlantic recolonization seems to have been from Europe to North America, rather than the other way (Wares and Cunningham, 2001).

Discusses the most recent published and unpublished work on which the study is based, showing that the proposal is based on existing work for which preliminary data are available and therefore giving credibility to the authors and strengthening the proposal.

Recent phylogeographic work

In a study of rough periwinkles, Wilding and co-authors (Wilding *et al.*, 2000a) describe a complex pattern of variation in mitochondrial cytochrome oxidase I. These authors considered that in *Littorina saxatilis* the complex picture of haplotype variation was probably to be resolved into two groups, one of which is now predominantly western and the other eastern on British mainland coasts. Similar features were seen in the closely related *L. arcana*, but this was not considered to be the result of hybridization: as in earlier work there was no evidence of hybridization in the field between these species (Ward, 1990). The third species in the group, *L. compressa*, is of too restricted distribution to be easily used for comparison.

Corroboration of large-scale differences in *Littorina saxatilis* on British shores comes from unpublished work by Wilding and Grahame. Here we have examined inter-population relationships using Nei's genetic distance based on mitochondrial CoI restriction fragment length polymorphism data, and compared the distance matrix obtained with geographical distances (see Figure 8.1). For a data set representing approximately the southern half of mainland Britain the expected isolation-by-distance pattern is not observed. When genetic and geographical distances are compared, the pattern is heterogeneous with a significant negative relationship.

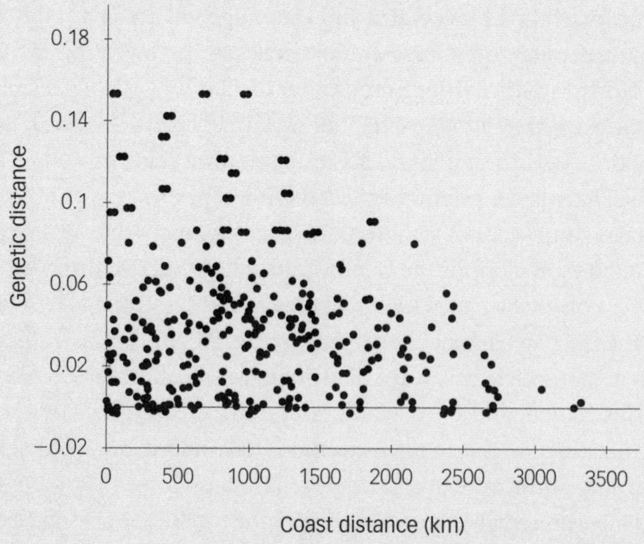

Figure supports the description of the unpublished data and the resulting conclusions

Figure 8.1 Genetic and coastal distances for 312 pairwise comparisons of sites on the British coast between St Abb's Head on the east coast, and Galloway on the west. There is a significant ($r = -0.159$, P 0.012) negative relationship (Mantel matrix test) between the distance matrices (Wilding and Grahame, unpublished).

There are two principal ways in which this pattern might have come about: the first involves invasion of the modern coastline starting from the south, with populations progressing from the Channel coast northwards along the east and west sides of the landmass. It is plausible that as successive waves of colonization moved northwards, the **same** rare alleles were independently lost on each colonizing front. This could result in the curious observation that populations on the south-eastern and south-western Scottish coasts (as far north as we have data for) are more alike than are many populations physically much closer together. However, there is an alternative mechanism: recolonization from two refugia, northern as well as southern, in which isolated populations had diverged.

In dog-whelks, Kirby and co-workers (Kirby and Bayne, 1994; Kirby *et al.*, 1997) have described a cline in shell shape, polymorphic enzymes, and mitochondrial DNA. In this last, a mutation in mitochondrial ND1 was found to vary in frequency with exposure. We (Case, Grahame and Wilding, unpublished) have examined populations on Anglesey (North Wales, west coast) and at Robin Hood's Bay (Yorkshire, east coast) determining shell shape in relation to exposure and looking for the same mutation. Shell shape variation conformed to expectation, with short-spired animals in exposed locations, and on Anglesey (Cable Bay; Menai Straits) the 'exposed shore G mutation' also behaved as expected, being more common in exposure. However, this was not true at Robin Hood's Bay. More interestingly, we found a mutation not reported by Kirby (Kirby *et al.*, 1997), which was present in both our east and west coast samples. Anglesey is at the south-western limit of a large 'eastern anomaly' among west coast dog-whelk populations (Crothers, 1985a), while the shores used by Kirby and co-workers are unambiguously 'western'. Perhaps, therefore, the Anglesey populations share an 'eastern' characteristic as would be predicted from the work of Crothers (e.g. Crothers, 1985b). This observation suggests that dog-whelks would be useful subjects for a molecular phylogeographic examination like the ones outlined for *Littorina saxatilis*, while at the same time we would wish to extend and deepen our understanding of the situation in *L. saxatilis*. We also wish to develop an entirely new set of observations using two topshell species, the rationale for which is explained below.

Overall, the background demonstrates a broad and detailed understanding of the subject area. The references cited are restricted to those which are essential to understanding the study. They are up to date and include unpublished data

Hypotheses

We suggest that northern European shores must have been recolonized from refugia in which members of the current biota survived the environmental upheaval of the last glaciation.

The background provides the rationale for the alternative hypotheses the investigator proposes

This recolonization may have been entirely from the south, in which case it was probably sourced by the present Iberian peninsula coasts. In this case we would predict:

1. Greater genetic diversity southwards, as one approaches the likely refugium, and genetic impoverishment northwards as one follows the route of recolonization.
2. Progressively greater accumulation of genetic distance comparing southern populations with northern ones.

It is also possible that there were northern refugia, so that the recolonization of British shores might have occurred from both the south and the north. In this case we would predict:

3. A less well-defined change in genetic diversity on the north–south gradient, although it is possible that it would still show an overall decline northward if northern refugium populations had been smaller.
4. Genetic distance would be greatest across the zone where colonizers from south and north met one another, and would decline on either side of this 'suture' towards the respective refugium zones.

As a third possibility, we suggest that there might be a combination of these scenarios—all species may not have behaved in the same way, even within the fairly restricted biotic assemblage of the rocky intertidal. We would predict:

5. That depending on the climatic tolerance of the individual species, there might be a combination of these phenomena observable in one assemblage.

Objectives are clearly and concisely stated and limited in number so that they can be realistically achieved within the 3-year time scale

Objectives of the study

Our objectives are to:

- Sample several species across a geographical range from Northern Norway to the Iberian Peninsula.
- Characterize the DNA polymorphisms within the sampled species.
- Analyse the polymorphism data to make inferences about the pattern of recolonization of northern European shores after the last ice age, with particular reference to the British coast.

The novel and significant features of the study are highlighted

Significance

This will represent the first study of the shoreline complementary to the body of work which now exists for the land surface, showing how populations responded to the climatic perturbation of the end of the last ice age. In particular it will allow tests of hypotheses concerning the distribution of refugia for different species. We believe that this work is timely given the current level of interest in the effects of climate change.

Experimental design

This section is structured, with a sub-section for each objective describing the appropriate experiments

Sample several species across a geographical range from Northern Norway to the Iberian Peninsula (1–18 months)

In the field, we shall need to sample a sufficient range of populations for the target species to describe genetic variation around Britain and in potential refugial areas. We seek to study a selection of animals which shows a good representation across the geographical range, in which there are contrasting reproductive modes (planktonic versus direct development) and with which we already have the necessary taxonomic expertise. These criteria are best met among the gastropod molluscs (snails) of the intertidal, and we propose to work with five species:

The samples which will be used are described and a rationale for their use is included

- *Littorina saxatilis*. We already know something about this in Britain, and the questions it suggests lie behind this proposal.
- *Nucella lapillus*. Again we know a little about this, though a lot less than about *L. saxatilis*; it shares with this species the habit of direct development (and therefore lacks a planktonic dispersal phase). Its morphometrics have been studied and suggest interesting questions; it also shows genetic polymorphism suggestive of interest. In Britain, the animal is considered to occur in eastern and western forms, and we have identified a mutation which fits with this hypothesis (Grahame, unpublished). These data are preliminary but suggest that further study would be worthwhile.
- *Littorina littorea*. This species is relatively unstudied. We include it because, in contrast with the first two, it has widespread planktonic dispersal, and this certainly seems to depress the potential for genetic variation over 200 km in Sweden (Janson, 1987). By analogy with other species, across a greater geographical range (as envisaged in this proposal) we might expect variation (Wares and Cunningham, 2001). Nevertheless, it is our best candidate for a widely distributed null model.
- *Trochocochlea (=Monodonta) lineata*, which has a planktonic dispersal phase, and reaches its northern limit on Anglesey (Lewis, 1964). This could be a particularly interesting species since it is

known to have suffered local extinction near its northern limit in the exceptionally cold winter of 1962/3, since when it has recolonized (Williamson and Kendall, 1981);

- Finally, *Gibbula umbilicalis*, which is like *Trochocochlea* in its breeding and distribution, although it extends its British range to the north coast of Scotland.

These species are eminently suitable for our purposes and we anticipate that they should allow us to distinguish between hypotheses 1 and 2 and 3 and 4 above, and indicate whether in fact the complex situation outlined in 5 is the case. They are all abundant, and there are no conservation implications arising from our proposed collection work.

The locations from which the samples will be collected are described. Again, a rationale is included

We propose to collect samples from 25 locations. These are:

- Three locations in Norway, in the vicinities of Tromsø, Trondheim, and Bergen, obtaining a spread of 1300 km. Final choice of location to be informed by consultation with contacts in Norway.
- 18 locations on mainland Britain, namely Duncansby Head, Peterhead, Arbroath, St Abb's Head, Durness, Applecross, Oban, Burrow Head (all in Scotland); Seahouses, Flamborough Head, vicinity of Cromer (for East Anglia), St Margaret's at Cliffe, St Alban's Head, Prawle Point, St Agnes (England, east and south coasts); St Ann's Head, Abersoch, and the Great Orme (Wales). The average 'direct sea distance' involved is about 160 km. An East Anglian site presents some problems due to the shortage of habitat; however, the four target species we would expect in this part of the range are all recorded in this area (Reid, 1996; Seaward, 1982).
- Two locations in France, in the vicinity of Cap de la Hague and St Nazaire (500 km sea distance).
- Two locations on the Iberian peninsula, in the vicinity of Santander and Lisbon (900 km sea distance), again final choice to be made after consultation with contacts in Spain/Portugal.

These locations have been chosen to give a spread of sampling sites within the ranges of our most widespread target species, though two of them reach their northern limits within this (see below). At each location we will make replicate samples taking 50 of each of our target species at each of two stations, the object of replication is to establish a framework of local variation. As well as the {(50 + 50) * target species} pattern for analysis, we will make a further two replicate samples of each species for cold storage in Leeds. These samples will be collected ~500 m from the first pair, which themselves would be best separated

Quantitative aspects of the study include sample size, number of species, and number of locations by ~100 m. Then in the event that we consider that further replication would have been desirable, the material for this will be available. Assuming that we analyse 30 individuals from each of the main replicated series, we envisage analysing a total of 5880 individual snails.

Characterize the DNA polymorphisms (1–30 months)

We will extract genomic DNA from the snails for analysis using standard protocols with which we are very experienced; spare material will be frozen (–70 °C).

We will screen for polymorphisms using single-stranded conformational polymorphism (SSCP) which is a relatively quick and inexpensive methodology for the screening of multiple samples for single point mutations (Hayashi, 1991). Because SSCP typically has a sensitivity of more than 99% (i.e. the probability of detecting at least one strand shifted) the vast majority of mutations will be picked up through simple gel analysis of treated PCR products, so reducing sequencing costs and allowing the screening of high numbers of samples. All detected variants can then be sequenced using standard sequencing methods thus estimating the frequency of known haplotypes. We have established sequencing facilities at the university which will support the sequencing of our samples.

> Detailed experimental descriptions are not included where procedures are standard (such as DNA extraction and SSCP). However, important details such as molecular markers and control genes are included

The molecular marker we will use to study variation is a section of mitochondrial Cytochrome-*B*. This gene has been used successfully to study variation in *Littorina* using SSCP (Small and Gosling, 2001) and in *Nucella* (Collins *et al.*, 1996). Another candidate with which we have experience is mitochondrial Cytochrome oxidase I. To allow comparison with a nuclear gene, we intend to use our knowledge of intron 3 from calmodulin in *Littorina* to use this gene in these animals (Wilding *et al.*, 2000b). For *Littorina* we have designed specific primers so that we can avoid amplifying from more than one copy. As the methodology is now established in our laboratory, we are confident of extending this to design specific primers for the other species.

> Overall, this study proposes the use of established protocols with which the investigators have experience and therefore the potential problems with the study are low

Data analysis (30–36 months)

We will analyse the data obtained using F_{ST} approaches (Hudson *et al.*, 1992) and the gene-tree population-tree coalescence software developed at the University of Arizona (Maddison and Maddison, 2000). As appropriate, we may use AMOVA (Excoffier *et al.* 1992) and genetic distance approaches. Applying such analyses to clearly framed hypotheses (Knowles, 2001) allows progress in answering the questions we have posed and hence to establish the pattern of recolonization of northern European shores after the last ice age.

> Specific methods that will be used to analyse the data are included, but details are not given as the methods are standard. Instead, the reader is referred to the original source of information

Timeline of the project:

Gannt chart showing different phases of the project

Objectives	Months					
	6	12	18	24	30	36
Collect samples	■	■	■			
Develop primers	■	■				
SSCP		■	■	■	■	
Cloning and sequencing		■	■	■	■	
Data analysis					■	■

The budget lists all the items required to complete the project and the costs associated with each. This budget is divided into three categories of costs (staff salaries, travel and subsistence, and consumable costs) for every year of the project. Following from the budget, detailed justification of each category of costs is provided

Budget and justification of the resources requested

Budget

Research salaries

Researcher 1

Person X (name anonymized)

Spine point 7, Grade R1A4

Age 25 (26 on proposed start date)

	Year 1	Year 2	Year 3	Total
Salary	£20 470.00	£21 503.00	£22 522.00	£64 495.00
London allowance—NA	£0.00	£0.00	£0.00	£0.00
National Insurance	£1 394.00	£1 486.00	£1 577.00	£4 457.00
Superannuation	£2 866.00	£3 010.00	£3 153.00	£9 029.00
Total	£24 730.00	£25 999.00	£27 250.00	£77 981.00
Researcher 2—NA	–	–	–	–
A Total research salaries	£24 730.00	£25 999.00	£27 250.00	£77 981.00
Associated costs				
Clerical/secretarial assistance	–	–	–	–
Technical assistance	–	–	–	–
Computing expenses	–	–	–	–
Travel/subsistence	£2 480.00	£5 000.00	£160.00	£7 640.00
Consumables	£6 490.00	£4 700.00	£4 380.00	£15 570.00
Other	–	–	–	–
B Total associated costs	£8 970.00	£9 700.00	£4 540.00	£23 210.00
Grand Total (A+B)	£33 700.00	£35 699.00	£31 790.00	£101 191.00

Justification of resources requested

The applicants justify the appointment of person X (name anonymized) for this project because of his demonstrated skill and expertise in molecular taxonomy of intertidal molluscs. In the course of his PhD work he has extracted DNA from all 19 species of the genus *Littorina*, usually working with ethanol-preserved material, and has successfully amplified the third intron of the calmodulin gene from these extracts. He has taken our understanding of the gene forward considerably: for example, we now understand that it is a multi-copy nuclear gene with up to five copies in the genome. He has presented papers on his work at several scientific meetings in the UK, and a poster at the European Society for Evolutionary Biology in Aarhus (2001). He is also a competent field worker, knowing the rocky shore and the species on it. We consider that he is an eminently suitable candidate to undertake the programme in this application.

Travel and subsistence

We envisage that field trips from Leeds to Durness (1000 miles), Stranraer (460 miles), Holyhead (330 miles), Penzance (806 miles), Dover (544 miles), and Dorchester (602 miles) (total mileage 3742, @ 25p per mile cost is £936) will be necessary to cover sampling sites in the UK. We have then budgeted for 13 days away to accomplish this, for two investigators at £45 per day subsistence (except for Holyhead), with a cost of £1125. Costs for collecting in Norway and the Iberian peninsula are based on consultation with travel companies and experience in the laboratory (colleagues have field sites in Spain).

Consumables

Breakdown of consumables

		Year 1	Year 2	Year 3	Totals
1	PCR	£1500	£1000	£ 500	£3000
2	Agarose gels for PCR evaluation	£ 400	£ 270	£ 130	£ 800
3	SSCP	£ 800	£ 500	£ 360	£1660
4	Sequencing	£1000	£ 900	£ 300	£2200
5	Cloning	£1020	£ 360	£1620	£3000
6	Extraction	£ 500	£ 400	£ 200	£1100
7	Gloves, lab plastic	£ 420	£ 420	£ 420	£1260
8	Maintenance lab facilities (incl. waste disposal)	£ 850	£ 850	£ 850	£2550
	Totals	£6490	£4700	£4380	£15570

1. **PCR techniques** will be used to amplify target sequences. This cost is based on 5880 amplifications @ 50p per reaction. We envisage that a high-grade taq polymerase will be necessary in order to minimize amplification error, given the sensitivity of the SSCP technique. The cost here also includes that of primers and dNTPs.
2. **Agarose gels for PCR evaluation.** As a matter of routine PCR products will be evaluated on agarose gels. This cost is based on experience in our laboratory.
3. **SSCP.** With this technique 30 individuals from each population will be screened for two markers giving an estimated total number of 2940 individuals, 5880 DNA samples. Allowing for an additional 20% contingency (inclusive of those lanes that will be used as standards) gives a total of approximately 7000 samples to be screened. Since there are 20 lanes per gel this means a total of 350 gels to run, total estimated cost £1660.
4. **Sequencing**. To maximize information, SSCP variants must be sequenced. This figure is very speculative and is based on the assumption that we may encounter 'regions' as follows, each of which will add SSCP variants:

Putative regions	Likely variant number	Species	Cost of sequencing
Norway	10	3	£ 300
Scotland	5	4	£ 200
England (E)	5	4	£ 200
England (W)	10	5	£ 500
Wales	5	5	£ 250
South coast	5	5	£ 250
France	5	5	£ 250
Iberia	5	5	£ 250
Total			£2200

5. **Cloning.** For most reliable sequencing, we will clone from PCR products representing new variants: this cost is based on experience in our laboratory bearing in mind the likely number of sequences.
6. **Extraction.** This cost is based on experience in our laboratory, bearing in mind the number of samples to be processed.
7. **Gloves, lab plastic.** A molecular laboratory demands use of gloves and sundry plastic disposables; this figure is based on experience.
8. **Maintenance lab facilities.** The School requires a contribution to be made to maintenance of laboratory facilities, including safe disposal of GMO material.

References

All the references cited in the main body of the text are included in the reference list and the referencing style used is consistent (author–date). References are used from a number of different peer-reviewed journals as well as books and an evolutionary analysis website. References date from 1964 through to 2001, with the majority of publications taken from the last 5 years, indicating that the material covered is thorough and up to date. (Note: this proposal was written at the start of 2002).

Butlin R. 1998. What do hybrid zones in general, and the *Chorthippus parallelus* zone in particular, tell us about speciation? In: Howard DJ and Berlocher CH, eds. *Endless Forms: Species and Speciation*. New York: Oxford University Press. 367–378.

Butlin RK, Ritchie MG, and Hewitt GM. 1991. Comparisons among morphological characters and between localities in the *Chorthippus parallelus* hybrid zone (Orthoptera: Acrididae). *Philosophical Transactions of the Royal Society of London. Series B: Biological Sciences* **334:** 297–308.

Cambridge PG, and Kitching JA. 1982. Shell shape in living and fossil (Norwich Crag) *Nucella lapillus* (L.) in relation to habitat. *Journal of Conchology* **31:** 31–38.

Collins TM, Frazer K, Palmer AR, Vermeij GJ, and Brown WM. 1996. Evolutionary history of northern hemisphere *Nucella* (Gastropoda, Muricidae): molecular, morphological, ecological and paleontological evidence. *Evolution* **50:** 2287–2304.

Crothers JH. 1985a. Dog-whelks: an introduction to the biology of *Nucella lapillus* (L). *Field Studies* **6:** 291–360.

Crothers JH. 1985b. Two different patterns of shell-shape variation in the dog-whelk *Nucella lapillus* (L). *Biological Journal of the Linnean Society* **25:** 339–353.

Crothers JH. 1992. A re-evaluation of shell shape in Shetland dog-whelks, *Nucella lapillus* (L.) and their use as biological exposure indicators. *Jounal of Molluscan Studies* **58:** 315–328.

Dawson AG. 1992. *Ice Age Earth: Late Quaternary Geology and Climate*. London: Routledge.

Excoffier L, Smouse PE, and Quattro JM. 1992. Analysis of molecular variance inferred from metric distances among DNA haplotypes: application to human mitochondrial DNA restriction data. *Genetics* **131:** 479–491.

Hayashi K. 1991. PCR-SSCP: a simple and sensitive method for detection of mutations in the genomic DNA. *PCR Methods and Applications* **1:** 34–38.

Hewitt GM. 1988. Hybrid zones—natural laboratories for evolutionary studies. *Trends in Ecology and Evolution* **3:** 158–166.

Hewitt GM. 1999. Post-glacial re-colonization of European biota. *Biological Journal of the Linnean Society* **68:** 87–112.

Hudson RR, Slatkin M, and Maddison P. 1992. Estimation of levels of gene flow from DNA sequence data. *Genetics* **132:** 583–589.

Janson K. 1987. Allozyme and shell variation in two marine snails (Littorina, Prosobranchia) with different dispersal abilities. *Biological Journal of the Linnean Society* **30:** 245–256.

Kirby RR, and Bayne BL. 1994. Phenotypic variation along a cline in allozyme and karyotype frequencies, and its relationship with habitat, in the dog-whelk *Nucella lapillus*, L. *Biological Journal of the Linnean Society* **53:** 255–275.

Kirby RR, Berry RJ, and Powers DA. 1997. Variation in mitochondrial DNA in a cline of allele frequencies and shell phenotype in the dog-whelk *Nucella lapillus* (L.). *Biological Journal of the Linnean Society* **62:** 299–312.

Knowles LL. 2001. Did the Pleistocene glaciations promote divergence? Tests of explicit refugial models in montane grasshoppers. *Molecular Ecology* **10:** 691–701.

Lewis JR. 1964. *The Ecology of Rocky Shores.* English Universities Press, London.

Maddison WP, and Maddison DR. 2000. Mesquite: a modular programming system for evolutionary analysis, version 0.98. *http://mesquite.biosci.arizona.edu/mesquite/mesquite.html.*

Pirazzoli PA. 1996. *Sea-Level Changes: the Last 20000 Years.* John Wiley and Sons, Chichester.

Reid DG. 1996. *Systematics and Evolution of Littorina.* The Ray Society, Dorchester.

Seaward DR ed. 1982. *Sea Area Atlas of the Marine Molluscs of Britain and Ireland.* Shrewsbury: Nature Conservancy Council and Conchological Society.

Small MP, and Gosling EM. 2001. Population genetics of a snail species complex in the British Isles: *Littorina saxatilis* (Olivi), *L. neglecta* Bean and *L. tenebrosa* (Montagu), using SSCP analysis of cytochrome-*B* gene fragments. *Jounal of Molluscan Studies* **67:** 69–80.

Stewart JR, and Lister AM. 2001. Cryptic northern refugia and the origins of the modern biota. *Trends in Ecology and Evolution* **16:** 608–613.

Ward RD. 1990. Biochemical genetic variation in the genus *Littorina*. *Hydrobiologia* **193:** 53–69.

Wares JP, and Cunningham CW. 2001. Phylogeography and historical ecology of the North Atlantic intertidal. *Evolution* **55:** 2455–2469.

Wilding CS, Grahame J, and Mill PJ. 2000a. Mitochondrial DNA CoI haplotype variation in sibling species of rough periwinkles. *Heredity* **85:** 62–74.

Wilding CS, Grahame J, and Mill PJ. 2000b. Nuclear DNA restriction site polymorphisms and the phylogeny and population structure of an intertidal snails species complex (*Littorina*). *Hereditas* **133:** 9–18.

Williamson P, and Kendall MA. 1981. Population age structure and growth of the trochid *Monodonta lineata* determined from shell rings. *Journal of the Marine Biological Association of the United Kingdom* **61:** 1011–1026.

References

Department of Health (2007a). *Human Tissue Act 2004 information* [online]. Available at: http://www.dh.gov.uk/en/Publichealth/Scientificdevelopmentgeneticsandbioethics/Tissue/Tissue-generalinformation/DH_4102169 (last accessed 28 April 2008).

Department of Health (2007b). *ACDP guidance* [online]. Available at: http://www.advisorybodies.doh.gov.uk/acdp/publications.htm#biologicalagents (last accessed 28 April 2008).

European Research Council (2008) *ERC grant schemes. Guide for applicants* [online]. http://erc.europa.eu/ (last accessed 28 November 2008).

Health and Safety Executive (2008). *GMOs – the law* [online]. Available at: http://www.hse.gov.uk/biosafety/gmo/law.htm (last accessed 28 April 2008).

Home Office (2008a). *Animals in scientific procedures* [online]. Available at: http://scienceandresearch.homeoffice.gov.uk/animal-research/ (last accessed 28 April 2008).

Home Office (2008b). *Drug laws and licensing* [online]. Available at: http://drugs.homeoffice.gov.uk/drugs-laws/licensing/ (last accessed 28 April 2008).

Joint Nature Conservation Committee (2008). *Conventions and legislations* [online]. Available at: http://www.jncc.gov.uk/page-1359 (last accessed 6 May 2008).

Medical Research Council (2007). *Applicants handbook 2007–08.* [online]. Available at: http://www.mrc.ac.uk/ApplyingforaGrant/ApplicantsHandbook/MRC004229 (last accessed 28 April 2008).

National Health and Medical Research Council (2008). *Peer review process for funding commencing in 2009* [online]. Available at: http://www.nhmrc.gov.au/ (last accessed 28 May 2008).

National Institutes of Health (2008). *NIH grant cycle: application to renewal. Part 5: research plan* [online]. Available at: http://www.niaid.nih.gov/ncn/grants/cycle/default.htm (last accessed 28 April 2008).

Natural Environment Research Council (2008). *Background and guidance notes for reviewer.* [online]. Available at: http://www.nerc.ac.uk/funding/application/referee/ (last accessed 28 April 2008).

National Science Foundation (2008). *Grant proposal guide* [online]. Available at: http://www.nsf.gov/funding/ (last accessed 10 June 2008).

World Medical Association (2004). *Declaration of Helsinki: Ethical principles for medical research involving human subjects* [online]. Available at: http://www.wma.net/e/policy/b3.htm (last accessed 28 April 2008).

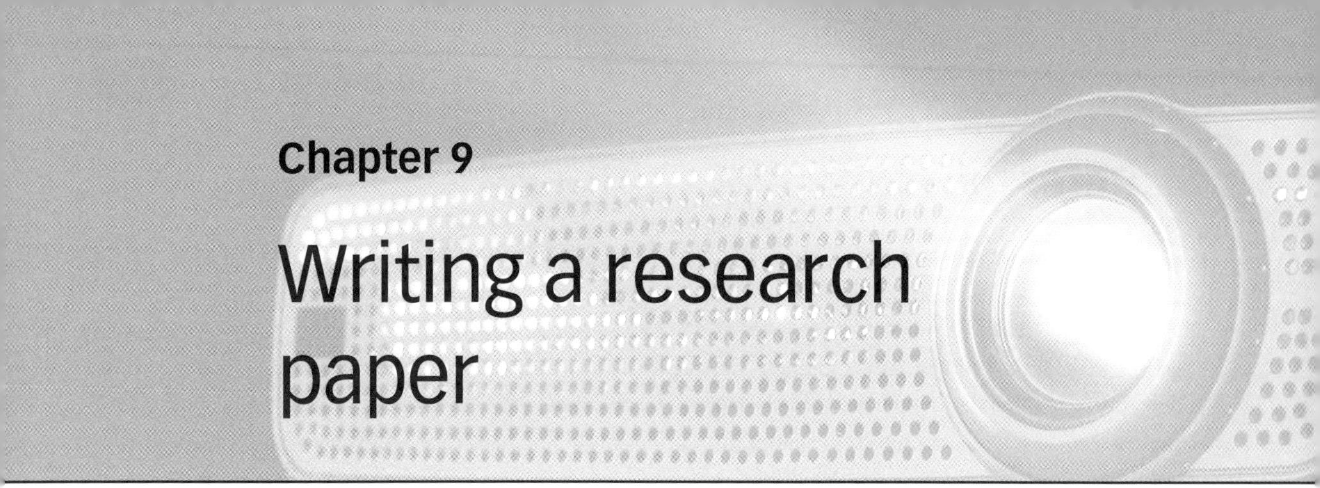

Chapter 9

Writing a research paper

➲ Introduction

Research papers are published in peer-reviewed academic journals and are an essential part of scientific communication as they report new information, which advances important conceptual or practical understanding of a particular problem or theory. These articles are detailed descriptions of the nature of the problem studied, the methodology used, the results achieved, and the conclusions reached.

The aim of this chapter is to guide you through the process of writing up your research findings in the form of a research paper for publication.

Specifically, this chapter will:

- identify the conventional components of the research article (title, authors, abstract, introduction, methods, results, discussion, references, acknowledgements, and supplementary information)
- describe a strategy for writing the research paper
- review the aim and typical content of each component of the paper
- include annotated examples of work to illustrate the content and structure of the various components.

9.1 What is a research paper?

The research paper is a primary source of literature (see Chapter 4, section 4.1) that reports original research which has not been published previously as a complete article. It is possible for the work (or parts of it) to have been presented at a conference in abstract, poster, or oral presentation format. Such formats do not constitute prior publication and therefore do not prevent you from publishing the full data in the form of a research paper in a peer-reviewed journal. However, if you have published substantial parts of the paper previously, this would be considered a duplicate publication (see Chapter 3, section 3.5.1) and therefore the paper cannot be resubmitted for publication.

The objective of writing and publishing a research paper is to communicate your research findings to the wider scientific community. By bringing the work into the public domain, it becomes part of the permanent scientific archive and allows other scientists to assess its quality and to build on it. Therefore, the written research paper must contain sufficient information about the research conducted in the laboratory or field to allow other scientists to:

- analyse and evaluate your findings (see Chapter 6)
- repeat your experiments if they wish to do so
- come up with new ideas to advance the area of research based on your findings.

9.2 Structure of a research paper

Research papers follow a basic structure in which the **I**ntroduction, **M**ethods, **R**esults and **D**iscussion (IMRAD—see Chapter 6) are presented sequentially. A variation to this structure is the one used by the journal *Cell* which positions the Methods section at the end of the paper following the Discussion. Other variations to the conventional IMRAD structure are to combine the methods and results under the sub-heading of 'Experimental' or to combine the Results and Discussion sections together under the sub-heading of 'Results and discussion'.

In addition to the Introduction, Methods, Results, and Discussion, a research paper includes additional components. These are typically:

- the title of the article
- authors
- keywords
- abstract
- acknowledgements
- references
- supplementary material.

The basic features of each of the components of a research paper are summarized in Table 9.1 and discussed in further detail in Section 9.4.

TABLE 9.1 The main sections of a research paper together with a brief description of the features associated with each section

Section	Brief description
Title	Short and descriptive; summarizes the content of the paper
Authors	Names, affiliations, and contact details of people who contributed to the work
Keywords	Words or phrases that are strongly associated with the content of the paper and are used by abstracting and indexing services to index the article
Abstract	A concise summary of the paper outlining the aims, the main methods, the main results, and the main conclusions of the study
Introduction	Provides a clear rationale for the study being undertaken by reviewing the published literature and describing clearly the aims/objectives of the study
Materials and methods	Outlines the materials and methods used to address the aims/objectives stated in the introduction
Results	A description of the main research findings illustrated with figures and/or tables.
Discussion	Contains a summary of the main conclusions, critically evaluates the results, and compares them with other published studies. Discusses the theoretical/practical implications of the work and makes recommendations for further work
Acknowledgements	Names of people or institutions that have assisted in the work (those not listed as co-authors) including details of funding sources
References	Lists all the references cited in the body of the text
Supplementary information	Typically supplied online and includes material that is beneficial (but not essential) to the reader in understanding the study.

9.3 Strategy for writing a research paper

Before starting to write a research paper you need to be sure that:

- you have sufficient data to write a full paper
- the data are original and important enough to warrant reporting.

If you are satisfied that your research findings fulfil both these criteria, then identify the journal you would like to publish in. Identifying your target journal before you start writing means you will be able to prepare your paper according to the specific instructions for authors provided by the journal of your choice. Remember, you will need to conform exactly to these guidelines. If you do not, the article will either be rejected or be returned to you for rewriting in the correct journal style. When selecting a journal to publish in, match the subject, novelty, and significance of your research to the type of articles your target journal publishes. These and other factors which may influence your choice of a journal were discussed in Chapter 4 (section 4.4) and you may want to go back and review this information before proceeding to the next section. You should also review the ethics of communication discussed in Chapter 3 at this stage: in particular, the ethical reporting of research data, copyright issues, plagiarism, duplicate submission and publication, conflicts of interest, and authorship issues.

Once you have familiarized yourself with the instructions, you can move on to the next step. The three stages involved in the writing process—planning, writing, and revising—are described below.

9.3.1 Planning the content of the paper

The first step is to define the scope of your paper. You can do this by writing brief summaries (approximately 50 words or less) to the following questions (adapted from Brown, 1994):

- What research questions have I investigated?
- Why was it important to study this?
- What are the results?
- What are the conclusions?
- What methods were used to obtain the results?
- What gap in knowledge did the work address?
- What is known at the end of the study that was not known before starting the study?
- What is still unknown about this area?

Another effective way of defining the scope of a paper is to structure your plan around a *central argument* (that is, the key claim) from which the rest of the paper components flow. When using this approach, first identify the key conclusions of your study and then the results that support these conclusions (see Chapter 6, section 6.3.1). The next logical step is to identify those methods that describe the results you intend to present. By the time you have done this, you will have defined the scope of the paper.

Next, you can begin to produce a more detailed outline of each component of the paper. Use the expected content of each component (described in section 9.4) to guide you in selecting the most appropriate information to include.

The more carefully you plan, the more likely your paper is to show a clear relationship between the goals of the study, the results presented, and the conclusions inferred from the findings. You are also more likely to avoid including irrelevant and extraneous material that could obscure the central focus of the paper. It is therefore worth spending some time on the planning stage of your writing.

9.3.2 Writing the paper

Once the planning is complete you are ready to start writing the paper. It is **not** a good idea to write it in a linear sequence, starting with the title first, working through each component of the paper in turn, and ending with the discussion. If you do this, you will almost certainly lose sight of the central arguments of your work. A more effective way of working is first to construct the figures and tables you will include as part of the Results section, and then write the text that describes them. Follow this up by writing the Materials and methods section of the paper. Then move on to write the Introduction, then the Discussion parts of the paper, and finally end with the abstract and the title. An alternative is to start with the Materials and methods, as this is straightforward, then write up the Results, the Discussion, and the Introduction, and again end with the abstract and the title.

Regardless of the order in which you write the sections of the paper, it is common practice to leave the abstract and the title till the end. Both these components must truly reflect the content of the paper, and this is only possible after the Introduction, Materials and methods, Results, and Discussion components are complete. If you do write the abstract or title at an earlier stage of the writing process, then go back and review them once the paper is complete to make sure that they are truly reflective of the content of the paper.

Finally, you can compile the remaining parts of the paper such as the supplementary material, acknowledgements, and authors' names and affiliations.

When writing a paper you must abide by the principles of accuracy, conciseness, clarity, and source referencing already discussed in Chapter 1, section 1.1 and summarized here:

- Report your work accurately. This includes using units of measurement, abbreviations, and scientific nomenclature (such as names of chemicals, genes, chromosomes, taxonomy) correctly and consistently. It also includes checking your figures and tables and any statements you make to see that there is no reason for anyone to misunderstand your work.
- Conduct up-to-date literature searches to make sure the information you are including is the most current.
- Write concisely and clearly. Use short sentences and avoid unnecessary repetition. Try to use simple language as far as possible. Of course there are times when technical words are necessary, but on the whole it is better to avoid long, complex terms. Concise and clear writing will make your work easier to follow and produce a shorter paper which will give lower production costs. This is important to journal editors as journals are usually sold for profit.
- When you use abbreviations, define them at the point of first use.
- An increasing number of journals prefer an active style of writing. However, check the conventions of the journal you intend to publish in, in case this is not their preferred style.
- Make clear what is your work and what is the work of others by referencing your work accurately and completely. This includes citing the reference within the text at the

point you refer to it and listing fully all the sources you have cited in the text at the end of your paper in the form of a reference list (see Chapter 1, section 1.2 and Chapter 3, section 3.4).

9.3.3 Revising the paper

Once the first draft of the paper is written, you should review the manuscript a number of times until you reach a final version you are satisfied with. A checklist for reviewing a research article is included in Box 9.1. In addition to this self-review, the paper will of course be improved by suggestions made by any co-author who reviews the manuscript. You may also want to ask another colleague not involved in the work but familiar with the subject matter to look through and comment on the paper: a fresh pair of eyes can usually pick up things that eyes that have viewed the same material several times will not spot.

BOX 9.1 Reviewing your research paper—checklist

Title page

❑ Does the title page include the article title, author names (with their addresses), and keywords?

❑ Does the title conform to the length limit stipulated in the instructions supplied?

❑ Is the title descriptive of the content of the article?

❑ Are there any superfluous words in the title that can be removed to improve conciseness and clarity?

❑ Are the authors' names and initials presented in the correct format requested by the journal?

❑ Are the keywords strongly associated with the content of the article and therefore suitable for indexing and retrieval purposes?

Abstract

❑ Does the abstract conform to the length limit stipulated in the instructions supplied?

❑ Does the abstract conform to the style stipulated in the instructions (i.e. structured abstract or single paragraph)?

❑ Does the abstract include the four elements: purpose, methodology, results, and conclusions, arranged in this order?

❑ Is the abstract free from references, illustrations, and unnecessary abbreviations?

❑ Are there any unnecessary words or sentences in the abstract that can be removed to improve conciseness and clarity?

❑ Can the abstract be understood without reference to any further information?

BOX 9.1 Cont'd

Introduction

- ❑ Does the Introduction provide an up-to-date account of the background to the problem you are studying?
- ❑ Does the Introduction provide a clear statement of the key objectives of the study?
- ❑ From the Introduction, is it clear which gap in the literature the study is addressing and why it is important to study it?
- ❑ Is the Introduction structured so that it starts by introducing the scope of the paper, moves on to summarize the current understanding of the problem, and ends with the key objectives of the study?

Materials and methods

- ❑ Are the subjects under investigation clearly identified?
- ❑ Are the experimental procedures described in sufficient detail to allow another competent scientist to repeat the experiments?
- ❑ Are quantitative aspects of the study included (e.g. volumes, quantities, concentrations, incubation times, etc.)?
- ❑ Are supplier locations for materials and specialist equipment included?
- ❑ Are methods of analysis described in sufficient detail?
- ❑ If standard methods have been used, have you supplied a reference?
- ❑ If modifications have been made to any standard procedures, have you described the modifications clearly?

Results—figures and tables

- ❑ Have you chosen the best way in which to summarize your numerical data (e.g. averages, percentages, ratios)?
- ❑ Have you chosen the best format in which to present your numerical data (e.g. tables or graphs)?
- ❑ Are all the illustrations necessary to understand your conclusions?
- ❑ Are there any illustrations that can be combined to produce a single multipart figure?
- ❑ Does each table have a number, title, and explanatory footnotes (as necessary)?
- ❑ Does each figure have a number, title, and a descriptive legend?
- ❑ Have you included standard deviations or standard errors where required?
- ❑ Are all figures and tables appropriately labelled (e.g. each column or axis) with the correct units of measurement where necessary?

- ❑ Have you explained any symbols, abbreviations, colour, or shading you may have used?
- ❑ Can each illustration be understood without reference to the main text of the article?
- ❑ If your figures are related, have you used symbols, shading, and line styles consistently?
- ❑ Have you saved the images in the correct format, resolution, colour mode, and size according to any instructions you have been supplied with?
- ❑ Will the figure and table information still be legible if it is reduced in size at publication?
- ❑ Have you obtained permission for any tables or figures you have reproduced for publication purposes?

Results—text-based descriptions

- ❑ Have you checked the accuracy of all the data you are reporting?
- ❑ Are all the figures and tables cited in the text?
- ❑ Have you described your key findings, trends, and comparisons in the text?
- ❑ Are these descriptions clear and concise or do they need further drafting?
- ❑ Have you backed up any qualitative statements such as 'markedly increased' with quantitative evidence?
- ❑ If you have used the word 'significant' have you backed it up with statistical evidence?

Discussion

- ❑ Are all the conclusions made supported by factual evidence?
- ❑ Have you considered all possible interpretations of the data?
- ❑ Have you compared your results with published literature and attempted to discuss how they support or contradict each other?
- ❑ Have you discussed the wider theoretical and/or practical implications of your work?
- ❑ Have you made recommendations for further work?
- ❑ Have you differentiated clearly between speculation and what is based on factual evidence?
- ❑ Is your discussion free from excessive speculation?
- ❑ Does the discussion link back to the questions posed in the Introduction?
- ❑ Does your discussion identify clearly how your work has added to the body of knowledge in the field you are working in?

BOX 9.1 Cont'd

Acknowledgements

- ❑ Have you acknowledged the assistance provided by people and institutions?
- ❑ Have you acknowledged all the sources of funding that supported your research?
- ❑ Does the acknowledgement format conform to the guidelines provided by the journal?

References

- ❑ Are your sources of information authoritative?
- ❑ Are your references relevant and up to date?
- ❑ Are all the sources you have used appropriately acknowledged in the text?
- ❑ Are all the sources cited in the text also listed at the end in the reference list?
- ❑ Have you used a consistent style of referencing both in the text and in compiling the reference list?
- ❑ Does the format of the references conform to any guidelines provided?

Supplementary material

- ❑ Is the supplementary material beneficial to the reader in understanding the study (without it being essential)?
- ❑ Is the supplementary material cited in the main text of the article?
- ❑ Have you checked the accuracy of this material?

Coherence and organization of your writing

- ❑ Have you used appropriate sub-headings to structure the article?
- ❑ Does each paragraph cover one topic only?
- ❑ Do the paragraphs follow in a logical manner so that the line of reasoning is clear?
- ❑ Are appropriate transitions used to link each successive paragraph?
- ❑ Are there any unclear or long sentences that can be rewritten to improve clarity?
- ❑ If you have used any abbreviations, symbols, or units are they in the correct format and consistent in style?
- ❑ Have you defined any abbreviations at first use?
- ❑ Have you checked for mistakes in spelling, grammar, and punctuation?
- ❑ Have you checked the layout of the document (e.g. font and size, spacing, margin size) to see that it conforms to any guidelines provided?

Submission of your manuscript for editorial assessment and peer review

- ❑ Are the text, figures, and tables in the format required by your target journal?

- ❑ Are all the appropriate forms and declarations included (conflict of interests, ethical approval information, co-author signatures)?
- ❑ Have you included a covering letter to the editor (if necessary)?

9.4 The aim and typical content of a research paper

To be able to prepare a good research paper you must be aware of the *aim* and *typical content* of each component of the paper. If you read scientific literature actively (see Chapter 6, section 6.3.1) and critically (section 6.3.2), you will be able to identify the good and not-so-good elements of a paper and use that information to guide your writing.

This section describes the aim and typical content of each component of a paper and uses annotated examples of published work to illustrate the points made. If you are unfamiliar with the subject matter used in the examples, do not worry: you should still be able to use the descriptive annotations to write and organize your work.

9.4.1 Writing the title

The aim of a title is two-fold. One is to inform the reader of the content of the article and the other is to help indexing and abstracting services to categorize the paper. Therefore, a well-written title should (1) concisely and accurately describe the essential content of the paper within the word limit stipulated by your target journal, (2) be informative so that it helps the reader decide whether they would like to read the full article or not, and (3) include key terms that will allow wide retrieval (and hence readership) by people searching the literature in the field you are writing in.

To illustrate appropriate and inappropriate titles, consider a paper reporting work in which a single point mutation is introduced into the gene that codes for the green fluorescent protein (GFP), which results in the amino acid tyrosine (Y) at position 66 being replaced by the amino acid tryptophan (W). The protein is then expressed and its fluorescent properties analysed. One possible title for this paper could be:

GFP properties

This is a very poor title: it contains an abbreviation and is too brief to reflect the essential content of the paper. An alternative might be:

Fluorescent properties of a mutant green fluorescent protein

This is better, as it is more descriptive of the content of the paper. However, an even better title is:

Y66W mutant produces green fluorescent protein with reduced fluorescence intensity

Although longer than the previous attempts, this is still concise as well as being descriptive of the study. In addition to meeting these two fundamental requirements of a title, it puts forward a specific conclusion. Some (but not all) journals accept (and indeed may prefer) this type of *declarative* title.

Title—some tips

- Make sure that the title is concise and accurately describes the essential content of the paper.
- Carefully select the words you include in the title. Use descriptive words that are associated with the content of your paper and therefore will be widely retrieved by online searches.
- Avoid starting titles with general words such as 'The . . .' or 'A . . .' and expressions such as 'Studies on . . .' or 'Observations of . . .'
- Avoid the use of abbreviations and non-essential words.

9.4.2 Listing the contributing authors and their addresses

The aim of listing the names of authors is twofold. One is to identify the people who have contributed to the work and therefore deserve credit for their input. The second is to identify the people who are taking public responsibility for the content of the manuscript. When listing co-authors there are two points to consider:

1. Who qualifies for authorship.
2. The order in which the authors are listed.

These points were discussed in Chapter 3 (section 3.5.3), to which you should refer. In addition to listing the names of authors who have contributed to the work, you should give the address of the institution to which each author is affiliated and provide the full mailing and e-mail address for the author to whom all correspondence relating to the article should be directed.

9.4.3 Selecting the keywords

The keywords should be terms that are closely related to the content of the article and can be used by abstracting and indexing services to index the article. It is common for journals to ask authors to supply a brief list of words or phrases (e.g. five key terms) that best characterize their study. You should select the keywords carefully so that they make your article easily retrievable by someone conducting an online search in the field of publication. Try to put yourself in their position. If you are using a new or relatively unusual method, it is worth listing this as a keyword as people working in different subjects but needing to use the technique will find your paper useful.

Examples of keywords for the study we described in section 9.4.1 could be **GFP**, **CFP**, and **mutation**.

9.4.4 Writing the abstract

The aim of the abstract is to convey the key content of the article to the reader in a concise manner. The abstract outlines the research aims, main methods, main results, and main conclusions of the study. Detailed guidance on writing an abstract is provided in Chapter 10.

9.4.5 Writing the Introduction

The aim of the Introduction is to provide a justification for the work you have undertaken. To do this, you need to review the published literature that relates to your study and clearly outline your research problem. Overall, the Introduction should set your work in context and convince the reader of the need to investigate the questions you have proposed. Chapter 7 describes how to write a critical review of the scientific literature which you should refer to as guidance for writing the Introduction to your research paper.

Briefly, an Introduction to a research paper:

- Starts with an introductory statement that briefly defines the scope of the work.
- Moves on to describe the current state of knowledge in the field. This includes a review of the recent and current developments that specifically relate to the topic under

Introduction—some tips

- Do not give an exhaustive review of all the literature that has been published previously. Instead, select the most important studies that are directly relevant to your investigation.
- Structure the Introduction so that it introduces the scope of the work and includes a summary of the key facts and theories that make up the topic area. Then move on to identify the gaps in the literature and end with a clear description of the objectives of the study. You may want to look through the annotated example of the background to the research proposal included in Chapter 8 (section 8.5) to see how the Introduction could be structured.

Introduction—some tips (Cont'd)

- If you contradict or reinforce some existing research in your Discussion, then make sure the relevant papers feature in the Introduction.
- Use the checklist provided in Box 9.1 to review the content and structure of the Introduction.

investigation. You should clearly establish what is known and what is unknown about the topic area, which specific gap in the literature you are addressing, and why it is important to study this.

- Concludes with a clear statement that describes your research question. This will include a brief description of the main objectives of the research as well as the approach taken to address the research question. Some papers may also include a brief summary of the key findings of the study.

9.4.6 Writing the Materials and methods section

The aim of the Materials and methods section is to describe your experiments in sufficient detail so that another competent scientist could repeat the experiments and verify your results (or at least assess the quality of your findings) (see Chapter 6). The Materials and methods section typically includes a description of the sample under study, the experimental procedures used, and how the data were analysed.

- **A description of the sample under study**. The samples may include human subjects, animals, plants, microorganisms, cell lines, or tissue samples. If using animals, plants, cell lines, or microorganisms you must state clearly which genus, species, or strain you are using and, for cell lines, whether it is a primary cell line or an established cell line. For a taxonomic study, it is important you show where the accession came from. For tissue samples, the source and how the material was obtained should clearly be identified. If the study used human subjects, you will need to describe how the samples were selected and include any relevant physical characteristics (e.g. age, sex). In any of these cases, if ethical approval was required to undertake the research, then details such as the name of the committee from whom ethical approval for the study was obtained should be included.
- **A description of the experimental procedures**. The amount of detail you include here will depend on whether the method you are describing is standard, or a published protocol, or a new method. If it is a published method then you do not need to provide detailed descriptions: instead, you can cite the publication in the text and list it as a reference. If your method is a standard protocol (such as streaking a bacterial plate or setting up and running an SDS–PAGE) then again you do not need to provide detailed descriptions, as people working in the field will be familiar with the methodology. However, if you have made modifications

to a published or standard protocol then these must be described. If your method is new, you will need to describe it in more detail. Irrespective of whether an experimental procedure is standard, published, or new, when writing up your methods you must include the precise details of the variable under study, the sample size, controls used, any treatments applied, and the number of times the experiment was repeated as well how the data were collected. If it is a field study, include a description of the physical and ecological characteristics of the study site, its exact location, and the environmental conditions under which the experiment was conducted (or observations made).

In addition, you must state precisely the quantitative details of your procedures. These could include, for example, the exact quantities and concentrations of any materials used, incubation times, temperatures, and equipment settings used to acquire data. It is also necessary to include the names of suppliers (and their locations) for materials (both biological and non-biological) and specialist equipment (e.g. microscopes) used during the study. This information is typically given in parentheses immediately after the description of the material or equipment. Supplier information is required so that scientists who wish to repeat the work can locate the same material and equipment types that you have used—this increases the likelihood of your data being independently reproduced. It is not necessary, however, to include routine equipment (such as pipettes, gel tanks, or ice buckets) when describing your methods as people working in the same field will be familiar with these.

- **A description of how the data were analysed**. This will include a brief description of how the numerical data were summarized (e.g. percentages, means, ratios), descriptions of any statistical tests used, and the name of any computer software used in the analysis of the results. In addition, if any images have been processed in any way, then details of the processing software used and the changes made should be clearly described (see Chapter 11, section 11.4). Again the level of detail you include will be determined by the novelty of the analyses. For example, if the statistical methods are standard or published, then do not include detailed descriptions but refer the reader to an appropriate reference (if one is necessary). However, if you are describing data analysis approaches that are new, then you should provide more extensive descriptions.

Annotated examples of well-written materials and methods

Below are three descriptions of materials and methods. The first is a description of a sample under study, the second is a description of experimental procedures, and the third is a description of how the data were analysed. These are extracts taken from a selection of published research articles to provide you with examples of how to write complete and succinct descriptions of the methods. Pay particular attention to the annotations which focus your attention on the types of information that should be included when writing the Materials and methods.

Example 9.1

Description of sample under study

(from Sugiura, M., Georgescu, N.M., Takahashi, M. (2007). *Plant Cell Physiol.*, **48**(7), 1022–1035, reprinted by permission of Oxford University Press).

Section is sub-divided

Name of the sample under study identified

Plant materials

Cucumber (*Cucumis sativus* L. cv. Jibai) seeds were imbibed in aerated water for 24 h and grown for 72 h in the dark at 28°C on vermiculite (GS; Nittai, Aichi, Japan) wetted with distilled water, unless otherwise indicated. Etiolated seedlings were then placed under continuous white light (120 μmol m^{-2} s^{-1}). Seeds of *A. thaliana* (L.) Heynh. Ecotype Columbia (Col-0) and corresponding T-DNA-tagged lines were provided by Dr. S Goto (Miyagi University of Education, Sendai, Japan) and the ABRC at Ohio State University (Columbus, OH, USA) respectively. The seeds were sown on rock wool (Airrich; Taiyo Kogyo, Tokyo, Japan) wetted with water, stratified for 3 d at 4 °C in the dark and cultivated on 1/1000-diluted Hyponex 6-10-5 medium (Hyponex Japan, Osaka) at 22°C under continuous light (180 μmol m^{-2} s^{-1}).

Describes how the samples were maintained during the experiment

Supplier names and locations included in parentheses

This description is well written because:

- The biological samples under study are identified by species and italicized according to convention (CSE, 2006) and include the taxonomic authority (in this case L. for Linnaeus).
- The conditions under which the samples were maintained is described concisely with the quantitative aspects clearly stated (e.g. temperatures, incubation times, wavelength of light used).
- The suppliers of biological samples are identified, with their locations.
- The suppliers of non-biological materials (e.g. rock wool) are also included.
- The section has sub-headings, for increased clarity.

Example 9.2

Description of experimental procedures

(from Ma, X., Fan, L., Meng, Y., Hou, Z., Mao, Y., Wang, W, Ding, W. (2007). *Mol. Hum. Reprod.*, **13**(8), 527–535, reprinted by permission of Oxford University Press).

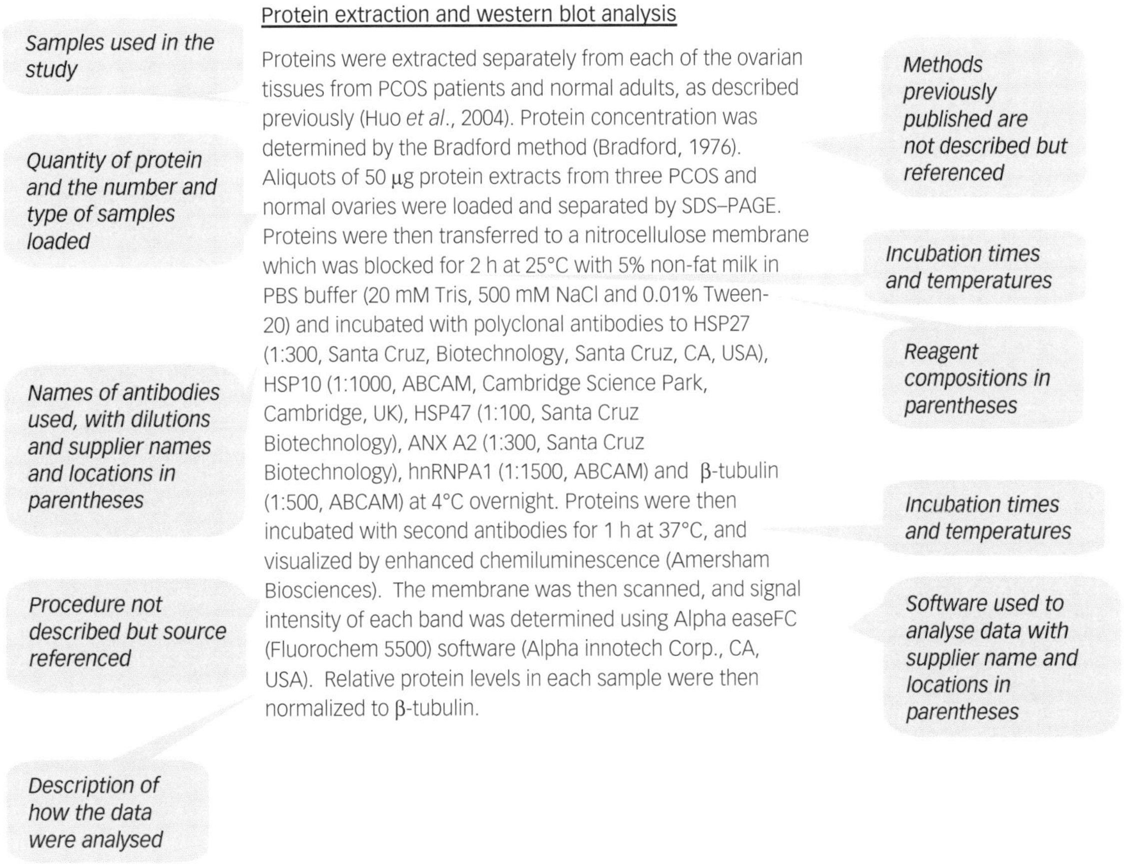

Samples used in the study
Quantity of protein and the number and type of samples loaded
Names of antibodies used, with dilutions and supplier names and locations in parentheses
Procedure not described but source referenced
Description of how the data were analysed
Protein extraction and western blot analysis
Proteins were extracted separately from each of the ovarian tissues from PCOS patients and normal adults, as described previously (Huo *et al*., 2004). Protein concentration was determined by the Bradford method (Bradford, 1976). Aliquots of 50 μg protein extracts from three PCOS and normal ovaries were loaded and separated by SDS–PAGE. Proteins were then transferred to a nitrocellulose membrane which was blocked for 2 h at 25°C with 5% non-fat milk in PBS buffer (20 mM Tris, 500 mM NaCl and 0.01% Tween-20) and incubated with polyclonal antibodies to HSP27 (1:300, Santa Cruz, Biotechnology, Santa Cruz, CA, USA), HSP10 (1:1000, ABCAM, Cambridge Science Park, Cambridge, UK), HSP47 (1:100, Santa Cruz Biotechnology), ANX A2 (1:300, Santa Cruz Biotechnology), hnRNPA1 (1:1500, ABCAM) and β-tubulin (1:500, ABCAM) at 4°C overnight. Proteins were then incubated with second antibodies for 1 h at 37°C, and visualized by enhanced chemiluminescence (Amersham Biosciences). The membrane was then scanned, and signal intensity of each band was determined using Alpha easeFC (Fluorochem 5500) software (Alpha innotech Corp., CA, USA). Relative protein levels in each sample were then normalized to β-tubulin.
Methods previously published are not described but referenced
Incubation times and temperatures
Reagent compositions in parentheses
Incubation times and temperatures
Software used to analyse data with supplier name and locations in parentheses

This description is well written because:

- The type of biological sample used in the experiment is identified (in this case ovarian tissues from patients with polycystic ovary syndrome (PCOS) and normal adult women).
- The quantitative aspects of the experimental procedure are clearly stated (e.g. amount of protein used, incubation times, antibody dilutions).
- Methods previously published are not repeated but the original publication is referred to (e.g. Bradford, 1976).
- Standard methods (such as SDS–PAGE) are not described.
- Supplier names and locations are included for reagents used (e.g. antibodies).
- Chemical compositions of reagents (i.e. PBS buffer) are identified in parentheses.
- The name and supplier of the software used for acquiring the data are included.
- The description has sub-headings, for clarity.

Example 9.3

Description of statistical analysis of data

(from Shi, Z., Zhang, H., Liu, Y., Xu, M., Dai, J. (2007). Alteration in gene expression and testosterone synthesis in the testes of male rats exposed to perfluorododecanoic acid. *Toxicol. Sci.*, **98**(1), 206–215, reprinted by permission of Oxford University Press).

Name of statistical software used, including supplier name and location in parentheses

Description of statistical tests used and details of how the data were analysed

Statistical significance level identified

Statistical analysis

All data were analysed using SPSS for Windows 13.0 Software (SPSS, Inc., Chicago, IL). All values are expressed as mean ± SEM. The ratio of testis organ to body weight was calculated to yield relative testis weights. Body weight and relative weight of testis were analysed using one-way ANOVA followed by Dunnett's *post hoc* two-sided *t*-test. Differences in testis weight, serum hormone concentrations, and gene expression levels between the treatment and control groups were analysed using a general linear model. Body weight was used as a covariant factor in analysis of these indicators. Dunnett's *post hoc* two-sided *t*-test was used to confirm differences between the control and treatment group. A probability (p) of less than 0.05 was chosen as the limit for statistical significance.

This description is well written because:

- The statistical tests used to analyse the data are described.
- The name of the statistical software used (with software version) is included, with the supplier's name and location.
- The statistical significance level is identified.
- The description has sub-headings, for clarity.

In all three examples presented above, the Materials and methods are concisely written without inclusion of any superfluous words. They all have subheadings, for increased clarity. In each case the descriptions are sufficiently detailed to allow the study to be replicated or at least allow the readers to assess whether the conclusions derived from the study are valid. This is exactly what the reviewers will be looking for when your manuscript is subjected to peer review, so keep this in mind when writing your Materials and methods.

Materials and methods section—some tips

- Include a description of the samples used, the experimental procedures, and how the data were analysed.

- Write succinctly, and state the quantitative components of the procedures precisely. It is a common mistake for new writers to provide long and wordy descriptions of methods which are lacking in precise quantitative detail.
- Describe only materials and methods that are necessary to understand the results you are presenting.
- Use sub-headings for clarity and match these sub-headings to those in the Results section so that the relationship between the two is apparent to the reader.
- Write the materials and methods section in the past tense. Although the examples presented here use the passive voice, the use of the active term 'we' is also acceptable. For example, 'we analysed the data . . .' instead of 'data were analysed . . .' However, check the conventions of the journal you intend to publish in.
- Order the sequence of methods to match the sequence in which the results are described.
- Use the checklist in Box 9.1 to review the descriptions of the materials and methods.

9.4.7 **Writing the Results**

The aim of the Results section is to describe your main research findings using a combination of text, tables, and figures. The figures and tables present the data while the text describes what the data are showing by summarizing the *key observations* and the *main trends*. When writing the Results section aim for accuracy, clarity, and simplicity. Use the minimum number of tables and figures required to understand your data and describe your observations as succinctly as possible.

The Results section is usually a *description* of your observations; you should leave the detailed interpretation of your findings for the Discussion section that follows. However, when describing your results it will be necessary to include at least some interpretation, so that the reader can see how one set of results leads on to the next. Some journals combine the results and discussions section together and, if you are writing for a journal which uses this format, then you will need to describe and interpret your results simultaneously.

Preparing the tables and figures

Make sure each figure and table is clear, accurate, and comprehensive enough to be understood without reference to the main body of the text. This means each table must include a title and explanatory footnotes, and each figure must include a descriptive legend. Chapter 11 provides guidance on presenting data in table or figure format.

Describing the figures and tables in the text

Data presented in table or figure format must be described in the text. These descriptions should concisely and clearly describe the key findings, make comparisons as appropriate, and highlight the main trends. The results-based text should not simply repeat information already presented in the figures and tables. Nor should it describe experimental procedures.

However, it is appropriate to provide indications of the methods used to contextualize the result without repeating the methods in detail (see Example 9.5 below).

Dividing the results into sub-sections and adding brief and informative sub-headings will add clarity to your results, particularly if the Results section is lengthy. When describing your data, make sure you cite the appropriate figure or table in the text. Tables and figures may be mentioned in parentheses, for example (Figure 1) or may be included in the main sentence in the format 'Table 1 shows . . . .' Whether you use the full word 'Figure' or the abbreviated form 'Fig.' will depend on the style used by your target journal. However, you should always make sure that you use a single consistent style throughout your article.

The units that you use to report the results will be stipulated by the instructions for authors. Most journals insist on the use of the SI system (see Chapter 1, section 1.1), but even so there may be differences between journals (e.g. use of cm^3, ml, or mL for volume). Ensure your units are in the correct format for your target journal.

Examples of textual descriptions of results

Below are two examples showing how data presented in an illustration can be described in the text. Each example is taken from a published article. Following each example is a summary of the good points of the text-based descriptions. Read through each carefully to see how descriptions of results can be written.

Example 9.4

Description of results shown in a table

This extract is taken from a paper which describes the effect of exposure to perfluorododecanoic acid (PFDoA) (a synthetic chemical) on male reproduction. Specifically, this extract describes the effect of PFDoA treatment on the body and testes weights of male rats.

The body and testes weights of male rats treated for 14 days with PFDoA are shown in Table 9.2. No significant changes in body weight were observed following exposure to PFDoA at 1 mg/kg/day compared with controls; however, for animals exposed to 5 and 10 mg/kg/day, body weight was significantly reduced by 25.7% and 38% respectively ($p < 0.01$). Though testes weight exhibited a degressive trend at 1 and 5 mg/kg/day, no significant differences were observed between the control group and these two treatment groups. In the rats receiving 10 mg PFDoA/kg/day, the testes weight was significantly diminished ($p < 0.05$). Relative weights were markedly increased at doses of 5 and 10 mg/kg/day ($p < 0.05$).

If you study the table and then read through the corresponding text-descriptions, you will see that the description is well written because:

- The text contains a clear and succinct description of the main findings presented in the table. The findings are not merely re-stated (e.g. the body weight at

TABLE 9.2 Body weight and testis weight of rats treated with PFDoA for 14 days

Doses (mg/kg/day)	0	1	5	10
Body weight (g)	292.30 ± 13.14	288.33 ± 10.55	218.70 ± 9.19**	184.50 ± 8.52**
Testis weight (g)	3.03 ± 0.18	2.74 ± 0.15	2.68 ± 0.19	2.35 ± 0.14*
Relative testis weight (%)[a]	0.97 ± 0.045	0.92 ± 0.056	1.32 ± 0.038**	1.32 ± 0.088**

Note. Data are given as mean ± SEM from 10 rats per group for weight or six rats per group for testis weight and relative testis weight.

[a] Percentage of total body weight.

Significant difference from control, * $p < 0.05$, ** $p < 0.01$.

(from Shi, Z., Zhang, H., Liu, Y., Xu, M., Dai, J. (2007). Alteration in gene expression and testosterone synthesis in the testes of male rats exposed to perfluorododecanoic acid. *Toxicol. Sci.*, **98**(1), 206–215. Extract and table reprinted by permission of Oxford University Press).

5 mg/kg/day is 218.70 ± 9.19 and the body weight at 10 mg/kg/day is 184.50 ± 8.52) but are actively compared (i.e. for animals exposed to 5 and 10 mg/kg/day, body weight was significantly reduced by 25.7% and 38% respectively). You should always avoid merely restating information presented in the tables but instead highlight the key trends and make comparisons.

- When the result is described as 'significant', the statement is backed up by statistical evidence. You should avoid the common mistake of using the word 'significant' without providing any statistical support. The use of this word in the context of describing your findings should mean that a statistical test has been conducted.
- Qualitative terms such as 'markedly increased' are backed up by numerical values. It is a common mistake for the author not to qualify terms such as 'there was an increase in . . .' and 'was lower than' and to leave the reader to work out what the extent of this increase or decrease is. When using such terms, always support them with quantitative evidence.
- A brief subtitle is included.

Example 9.5

Description of results shown in a figure

This extract is taken from a paper which describes the first confirmed nitrite transporter, CsNitr1-L, in the inner envelope of higher-plant chloroplasts. Specifically, this extract presents results which show that cucumber seedlings treated with light induce the expression of the nitrite transporter gene (*CsNitr1-L*) and its isoform (*CsNitr1-S*).

Northern blot analysis revealed light-induced expression of a transcript in cucumber seedlings (Figure 9.1a). The size of this transcript, about 2100 nucleotides, agrees with the size of cloned *CsNitr1-L* cDNA (1947 bp including a 20 bp poly(A) tail). There was no visible signal from this transcript in the etiolated seedlings before illumination was applied (–6 h and 0 h). After 6 h illumination, a positive signal appeared in the greening seedlings. The mRNA accumulated with continued illumination and seemed to reach a plateau after 36 h. At 60 h after illumination was first applied, the cotyledons had almost completed their expansion and greening, and had a Chl content of 100 μg (mg protein)$^{-1}$. *CsNitr1-S* was cloned from the cDNA libraries constructed from mRNAs of 6 h and 12 h illuminated cucumber seedlings. At 6 h illumination there were no clear separate bands. *CsNitr1-S* is transcribed at too low a level in the seedlings to be detected by standard northern blotting. Reverse transcription–PCR (RT–PCR) was conducted to determine separately the expression of *CsNitr1* transcripts during the greening of cucumber seedlings. RT–PCR amplifying the 5'-terminal sequence using mRNAs of 6 h light-grown seedlings gave two bands with a difference of about 220 bp in length (Figure 9.1b).

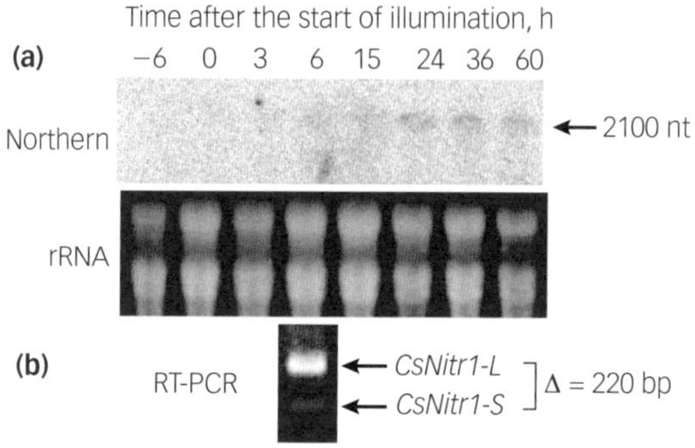

Figure 9.1 Light-induced expression of *CsNitr1*s in cucumber seedlings. (a) Total RNA (20 μg) of cucumber seedlings sampled after the indicated durations of illumination were subjected to northern blot analysis as described in the text. Membranes were exposed to imaging plates for 3 d and read by using a Fuji Film FLA3000 imaging analyser. A gel stained with ethidium bromide is shown in the bottom panel (labelled rRNA). (b) RT–PCR products of *CsNitr1-L* and *CsNitr1-S* in cotyledons after the indicated durations of illumination. The primers used were 5′-GCAAAGAGGTAAATAAGAATGG-3′ and 5′-TTTCGGTTCTTGAAGGCCGC-3′, corresponding to bases 1–22 and bases 836–817 in the *CsNitr1-L* sequence, respectively. A 10 μl aliquot of the PCR product (35 cycles) was run in a 2% agarose gel and stained with ethidium bromide. (from Sugiura, M., Georgescu, N. M., Takahashi, M. (2007). Light-dependent expression of *CsNitr1-L* in leaves. *Plant Cell Physiol.*, **48**(7), 1022–1035. Extract and figure reprinted by permission of Oxford University Press).

If you study the figure and read through the corresponding description in the text, you will notice that the description is well written because:

- The text contains a clear and succinct description of the main findings presented in the figure. Note that the description does not merely restate the findings (e.g. the size of the transcript on the northern blot is 2100 nucleotides and the size of the RT–PCR transcripts differs by 220 nucleotides) but instead makes comparisons (i.e. the size of this transcript, about 2100 nucleotides, agrees with the size of cloned *CsNitr1-L* cDNA (1947bp including a 20 bp poly(A) tail).

- Sufficient methodological detail is included to place the result in context (e.g. Northern blot analysis revealed . . . . Reverse transcription–PCR (RT–PCR) was conducted to determine separately the expression of . . . .) without repeating experimental procedures at length.
- Each part of the figure is cited in the text sequentially.
- A brief and informative subtitle is included which adds clarity to the results.

After reading through the two examples presented above you should be able to write your text-based descriptions in a way that uses the good practice identified here and avoids common mistakes.

Results section—some tips

- Sub-divide your results into sections and use descriptive sub-headings for clarity.
- Arrange your results logically so that one finding follows from another and leads the reader through your data in a way that answers the questions you have stated in your Introduction.
- Ensure that for every set of results you describe there is a corresponding method in the Materials and methods section of the paper explaining how the results were generated.
- Carefully select the data you will present in the form of figures or tables. Choose only the most essential evidence that answers the questions you have posed in your Introduction. This includes data that may not support your hypotheses.
- Make sure each figure or table is numbered, with a brief title and a descriptive legend that is self-explanatory so that the figure or table can be understood without reading the main body of the text.
- Describe the results shown in the figures and tables clearly and concisely without unnecessarily duplicating information. Use the examples provided above to guide you in how to write clear and concise descriptions.
- It is important that the reader can understand from the Materials and methods and the Results sections what statistical analyses were carried out and what the results of these analyses are. Therefore, make sure you include the outcome of the statistical tests in the Results section.
- Review the Results section of the paper using the checklist in Box 9.1.

9.4.8 Writing the Discussion

The aim of the Discussion is to evaluate your results critically and explain how your findings fit with the primary research work that has been published. To achieve this you will need to:

- Draw together the different strands of your results to provide a composite picture of how they can be interpreted.
- Compare your results with the results of other published studies to see how they support or conflict with your findings.
- Consider the wider implications (theoretical and/or practical) of your work.
- Make recommendations for further work.

The Discussion usually begins with a brief summary of the major findings before moving on to a critical assessment of the results and how they compare with other studies. It then ends with a concluding paragraph which emphasizes the main findings and makes recommendations for further work. These three parts of the Discussion are described in detail below.

- **Introductory paragraph**. It is common to start the Discussion by very briefly summarizing the major findings. This brief summary orientates the reader to the interpretation that is to follow. This first paragraph should focus on the key results only.
- **Main body of the Discussion**. This is the main and substantive component of your Discussion in which you present a coherent argument to the reader by leading them systematically through your line of reasoning to show the most plausible conclusions that can be inferred from your findings. As you write your Discussion:
  - Make sure you address the questions you have posed in your Introduction, so that the reader can follow the extent to which your research aims have been met. (Writing the Introduction after the Discussion will ensure that you have done this.)
  - Be critical in your interpretation of the results. Consider all possible conclusions that can be inferred from your data.
  - Compare your findings with existing primary published literature and discuss the extent to which they support or conflict with your work.
  - If there are limitations to your study, then discuss them. If it is difficult to reach a particular conclusion on the basis of the factual evidence available, then say so.
  - As research papers report original work and advance an existing body of knowledge you will need to highlight how your work is 'new'. This may involve proposing models that build on (or are an alternative to) an existing model. It may involve identifying how the work can be used practically.
  - Avoid common mistakes such as stating conclusions that are not supported by factual evidence and speculating excessively. Some speculation is, of course,

acceptable—for example, when proposing models. However, always differentiate clearly between what is speculation and what is supported by factual evidence.

- **Concluding paragraph.** The final paragraph of the Discussion should emphasize your main conclusions and the wider implications of your work. You should also make recommendations for further work here, if they have not been integrated into the main body of the Discussion.

Discussion section—some tips

- The Discussion section is the most difficult part of the paper to write and must be planned carefully. To assist you, use the planning techniques described in Chapter 1, section 1.3. You should also critically read a selection of papers in your subject area to see how the authors develop their arguments as they progress through the discussion.
- Use words such as 'therefore', 'so', 'as a consequence', or 'in conclusion' to signal to the reader that a main point is about to be made.
- Be critical in your interpretation of your results: identify all the alternative ways in which your data could be interpreted.
- Discuss how your data compare with existing published literature.
- Identify clearly what was unknown before the study and what is now known as a consequence of your work.
- Link your findings back to the research aims stated in the Introduction.
- Divide your Discussion into sub-sections if necessary for clarity.
- Use illustrations to explain difficult concepts (e.g. models) if appropriate.
- Use the checklist in Box 9.1 to review your Discussion.

9.4.9 Acknowledgements

The acknowledgement section follows the discussion and precedes the reference list. It consists of a short paragraph in which the authors acknowledge the assistance provided by any people or institutions that are not listed as co-authors. This will include those that have supplied you with any material (for example, an organism or a starting plasmid construct) or provided any technical assistance (see Chapter 3, section 3.5.4). In particular, you should acknowledge the source of funding that supported your research.

Example 9.6 shows an annotated example of an acknowledgement statement.

Example 9.6

Acknowledgements

We thank Mrs Jones for the technical assistance provided in analysing the flow cytometry data and Dr Klass in sharing the T-cell gene expression profile data set. This work has benefited from useful discussions with Dr Lima. The work was supported by Cancer Research UK and the Yorkshire Cancer Research Campaign.

Assistance provided by people acknowledged

Funding sources acknowledged

When writing the acknowledgements, check the instructions for authors of your journal of choice. Some journals may ask that details of all funding sources are provided separately from other acknowledgements, in a sub-section entitled 'Funding'. Some may ask specifically for grant and contribution numbers to be included but others may not (as in the above example).

9.4.10 **References**

As discussed in Chapter 1, section 1.2, there are three main reasons for referencing the sources of information you use in your work. One is to credit the original author whose work you are using, the second is to help identify and locate the source of information, and the third is to differentiate clearly which part of the work is yours (and therefore new) and which is the contribution of others. The key points to consider when referencing are:

- **How many references to cite**? The number of references cited in any particular paper will vary depending on the length of the article, but can range from approximately 25 citations to over 50. Some journals will specify the maximum number of references they will accept. For example, the journal *Nature* stipulates that no more than 50 references are used. Other journals may not stipulate exactly, but you could look through a selection of papers published in your target journal to get an idea of the number of references used by various authors.
- **Which references to cite**? When selecting your references, limit the list to those that are essential for understanding the scope of your work. You should as far as possible cite primary published literature, but where a comprehensive review article is available, citing this may be preferable to listing many separate references.
- **How to cite the references**? Always refer to the instructions for authors for the style of referencing used by the journal of your choice. These instructions will clearly outline (and in some cased provide examples of) how the references should be cited within the body of the text (numerical or author–date system) and how the complete reference should be listed in the reference list at the end of the document. If you use

reference management software to compile references, you will be able to reformat your reference list to conform to the journal style requested by the journal relatively easily (see Chapter 1, section 1.2).

9.4.11 Supplementary information

Information that is not essential for understanding your data and their interpretation, but which would nonetheless benefit the reader can be supplied as supplementary material for some journals. This additional material must be (1) directly relevant to your study (and can include information such as additional methodological details, additional figures, and data sets); (2) cited in the main text of the article; and (3) submitted at the same time as the main paper. This supplementary information will be made available online to the reader at the time of publication or, in some journals such as *Plant Physiology*, included in the published paper itself.

9.5 Submitting your completed manuscript

Once the final draft of your manuscript is complete you are ready to submit it to your target journal for editorial assessment and peer review. The instructions for authors for your target journal will provide very specific guidelines on how the paper should be submitted. These guidelines will relate to:

- **Online or hard copy submission**. Some journals will only accept online submissions of the prepared manuscript; others provide the option of online or hard copy submission. The advantage of using the online submission service is that the manuscript is handled electronically throughout the editorial assessment and peer-review process and therefore the time between submission of manuscript and acceptance is speeded up significantly.

  Electronic submission also has the added advantage of allowing the authors to track the status of the submitted manuscript and you can communicate with journal staff via e-mail links built into the online submission system.

  If you submit a manuscript electronically, the journal website will very clearly guide you through the submission stages. If you submit the manuscript as a hard copy, then you will need to make sure that the required numbers of copies of the manuscript are sent as stipulated by the journal.

- **The format in which the text should be submitted**. Most journals will ask that the text of the paper is submitted as either a word-processed file (.doc or .rtf), Adobe Acrobat (.pdf), or Postscript (.eps) file: but consult the instructions for authors.
- **The format in which figures should be prepared and submitted**. Each journal will provide very specific information on how figures should be prepared and submitted. These instructions will specify file formats in which the figures should be saved as well as figure resolutions, and whether the figures and tables should be embedded

into the main manuscript or supplied separately. Chapter 11 provides guidance on how figures should be prepared.

- **Completion of forms and declarations**. You will be required to disclose any financial or other interests that could be construed as conflicts of interest at the time of manuscript submission (see Chapter 3, section 3.5.2). You will also be required to complete an authorship form and supply any information relating to ethical approval (if required) at the same time.
- **Covering letter to the editor**. When submitting the manuscript, you will need to include a covering letter the editor. This letter should be brief and include the following information:
  - the title of the manuscript and the category or section of the journal the paper falls into (e.g. scientific paper or letter)
  - a few sentences about the content and significance of the paper
  - the names of all the authors
  - the name and address of the corresponding author
  - a statement to the effect that the paper has not been published previously nor has it been submitted elsewhere simultaneously for publication
  - a statement to say that you agree to pay the specified charges on publication
  - name and contact details of a number of potential reviewers and the names of those that the authors would prefer not to be used (such as researchers whose work may be directly related to the content of the paper) (see Chapter 4, section 4.2.2).

If you are submitting online, the system may provide on-screen instructions asking you to enter the above details directly onto the screen. For examples of well-written covering letters see Fraser (1997), Hills (1999), and Day and Gastel (2006).

9.6 Peer review and publication

Once the paper has been submitted to the journal it will undergo the peer-review process (see Chapter 4, section 4.2). If the outcome of the review is to accept the paper but with revisions, read through the reviewers' comments carefully to see what is required before the paper will be accepted for publication. Examples of the types of revisions that could be requested are reformatting a particular figure, or rewriting parts of the text such as the Introduction, or conducting further experiments. You do not have to accept all of the reviewers' suggestions. If you disagree with aspects of the reviewers' comments, then you can rebut with reasoned arguments stating why the suggestions have not been implemented. However, most papers are improved by taking account of the comments of the reviewers, so you should consider their suggestions carefully.

Sometimes a journal may reject the paper in its current form but encourage the author(s) to resubmit the manuscript after major revisions. Major revisions typically involve much more extensive experimentation.

If your paper is rejected, without encouraging you to resubmit, then one reason could be that the novelty and significance of your article do not match those of the type of articles the journal publishes. In this case you will need to consider an alternative journal, possibly with lower ranking, to publish your work in. Remember, if you resubmit your article to a different journal, then you will need to edit the paper in line with the style of that particular journal.

Another reason for rejection could be that your experiments are poorly designed or your analysis and interpretation of your data is poor (hopefully your paper will not be rejected on this basis!). In this case you should consider how to improve the quality of your work before resubmitting it to another journal.

When you receive your rejection letter (and most if not all scientists will have had papers rejected at some time in their career), you should consider the reasons for rejection carefully and then modify your paper accordingly. Remember, you work is less likely to be rejected if your experimental design, data analysis, and interpretation are of a high quality and you communicate this information in the form of a well-written and well-structured paper. Your paper is also less likely to be rejected if you are realistic in your choice of journal and hence match the novelty, significance, and subject area to the type of articles the target journal publishes.

References

Brown, R. (1994). The big picture about managing writing. In: Zuber-Skerritt, O. and Ryan, Y. ed. *Quality in postgraduate education*. Kogan Page, London.

CSE (2006). *Scientific style and format: the CSE manual for authors, editors, and publishers,* 7th edn. Council of Science Editors in cooperation with The Rockefeller University Press, Reston, VA.

Day, R.A. and Gastel, B. (2006). *How to write and publish a scientific paper*, 6th edn. Cambridge University Press, Cambridge.

Fraser, J. (1997). *How to publish in biomedicine*. Radcliffe Medical Press, Abingdon.

Hills, P.J. (ed) (1999). *Publish or perish*, 2nd edn. Peter Francis Publishers, Dereham, Norfolk.

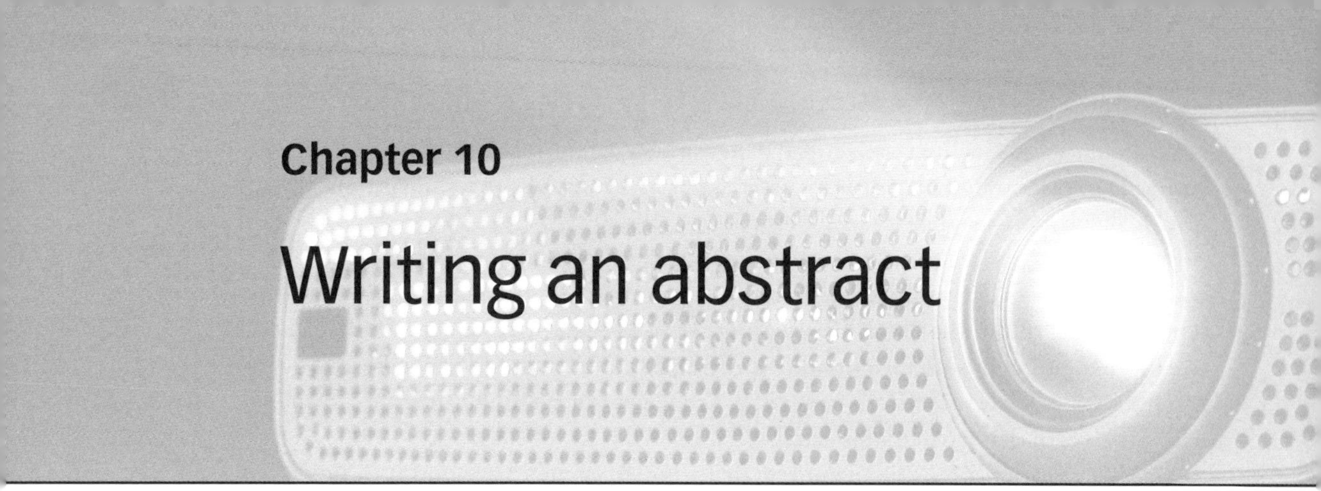

Chapter 10

Writing an abstract

➲ Introduction

An abstract is a short paragraph which summarizes concisely the content of your work. Abstracts form part of many types of scientific communications including research papers reporting original work, review articles, dissertations and theses, research proposals, and conference proceedings. The aim of this chapter is to develop your ability to write concise and comprehensive abstracts of your research findings.

Specifically, this chapter will:

- describe the purpose and essential features of abstracts
- provide annotated examples of abstracts
- provide tips on how to write a good abstract
- provide a checklist for reviewing an abstract.

10.1 **What is an abstract?**

An abstract is a short written summary, usually 200–300 words in length. Abstracts can form part of a longer communication such as a research paper, a review article, a research proposal, a dissertation, or a thesis, or it can be a stand-alone unit such as the abstract that forms part of a conference presentation. In all of these communications, the aim of the abstract is to convey the key content of the article or presentation (oral or poster) to the reader in a concise manner.

When the abstract forms part of a longer article, it is generally used by readers to decide if the work is relevant to their interest and if it is worthwhile reading the entire article or not. Remember that abstracts are widely available from literature searching programs and websites (see Chapter 5) and for many people the abstract will be their first introduction (and sometimes their only exposure) to your work. You should therefore write the abstract with care so that it conveys the message of your work clearly and entices the reader to read the full article.

A conference abstract is used by the conference committee to decide whether or not to accept your work for oral or poster presentation at the conference. It is therefore very important to write it in as concise and informative a way as possible so that the relevance and quality of your work can be judged properly by the readers.

All abstracts share common features:

- The abstract is written so that it stands alone—that is, the information is comprehensible to the reader without having to refer to the main article, poster, or oral presentation.
- It does not cite references.
- It does not include figures or tables.
- It keeps abbreviations to an absolute minimum.
- It is written (usually) in the active voice.

10.2 The research paper abstract

In a research paper the abstract appears before the main body of the article, after the title, authors, and keywords. It is the first (and often only) part of the paper, apart from the title, that a potential reader will look at. The content must therefore convey the key messages of your paper and be comprehensible to the reader without reference to the main article. An abstract typically contains four types of information, arranged in the following order:

- An introductory statement that identifies the purpose (aim) of the study. This is typically (but not always) preceded by an introductory statement that establishes the context of the study.
- The key methods utilized in the study.
- The main results of the study.
- The major conclusions of the study.

Example 10.1

Single-paragraph abstract

Here is an example of an abstract that has been annotated to highlight its key content and structure.

If you read through the abstract and the annotations you will see that the abstract is well written because:

- It includes the essential content expected in an abstract—that is, it establishes the context of the study and states the aims, the methods, the main results, and the main conclusions of the work.

An introductory statement that establishes the context of the study

Clear aim of the study

Experimental approach outlined briefly

Main results of the study identified

Abstract finishes with the main conclusion

Many animals possess camouflage markings that reduce the risk of detection by visually hunting predators. A key aspect of camouflage involves mimicking the background against which the animal is viewed. However, most animals experience a wide variety of backgrounds and cannot change their external appearance to match each selectively. We investigate whether such animals should adopt camouflage specialized with respect to one background or adopt a compromise between the attributes of multiple backgrounds. We do this using a model consisting of predators that hunt prey in patches of two different types, where prey adopt the camouflage that minimizes individual risk of predation. We show that the optimal strategy of the prey is affected by a number of factors, including the relative frequencies of the patch types, the travel time of predators between patches, the mean prey number in each patch type, and the trade-off function between the levels of crypsis in the patch types. We find evidence that both specialist and compromise strategies of prey camouflage are favored under different model parameters, indicating that optimal concealment may not be as straightforward as previously thought.

FIGURE 10.1 (from Houston, A.I., Stevens, M., Cuthill, I.C. (2007). Animal camouflage: compromise or specialize in a 2 patch-type environment. *Behav. Ecol.,* **18**(4), 769–775, reprinted by permission from Oxford University Press)

- The content is arranged logically, in the prescribed order: aims, followed by methods, then results, and ending with conclusions.
- The abstract is comprehensive enough to be understood without reference to the main article.
- It avoids abbreviations, references, and illustrations.
- It is written in an active voice (e.g. 'we find') which makes it easier and more enjoyable to read.

A variation of the single-paragraph abstract format described above is the structured abstract, in which the abstract text is divided into sub-sections under specific headings. These sub-headings can vary, but typically include Background, Methods, Results, and Conclusions. Examples of journals which use the structured abstract format are *BMC Biotechnology*, *JAMA*, *Bioinformatics*, and the *Journal of the Science of Food and Agriculture*.

Example 10.2

Structured abstract

Below is an annotated example of a structured abstract published in the journal *BMC Biotechnology*. This particular journal structures the abstract under three

Background
Human cell lines are the most innovative choice of host cell for production of biopharmaceuticals since they allow for authentic posttranslational modification of therapeutic proteins. We present a new method for generating high and stable protein expressing cell lines based on human amniocytes without the requirement of antibiotic selection.

Background establishes the context of the study and presents the aim of the work

Results
Primary amniocytes from routine amniocentesis samples can be efficiently transformed with adenoviral functions resulting in stable human cell lines. Cotransfection of the primary human amniocytes with a plasmid expressing adenoviral E1 functions plus a second plasmid containing a gene of interest resulted in permanent cell lines expressing up to 30 pg/cell/day of a fully glycosylated and sialylated protein. Expression of the gene of interest is very stable for more than 90 passages and, importantly, was achieved in the absence of any antibiotic selection.

The main results of the study are included. From the description of the results, the experimental approach used is also clear

Conclusion
We describe an improved method for developing high protein expressing stable human cell lines. These cell lines are of non-tumor origin, they are immortalized by a function not oncogenic in human and they are from an ethically accepted and easily accessible cell source. Since the cells can be easily adapted to growth in serum-free and chemically defined medium they fulfill the requirements of biopharmaceutical production processes.

The abstract ends with the main conclusions of the study

FIGURE 10.2 (from Schiedner, G., Hertel, S., Bialek, C., Kewes, H., Waschutza, G., Volpers, C. (2008). Efficient and reproducible generation of high expressing, stable human cell lines without need for antibiotic selection. *BMC Biotechnol.*, **8,** 13, reprinted by permission under the Creative Commons Attribution 2.0 license).

headings: Background, Results, and Conclusion. Overall, the content of the abstract includes four elements—aims, methodology, results, and conclusions—arranged sequentially.

10.3 The conference abstract

If you want to present your initial findings at a scientific conference then you will need to submit an abstract of your research work. This abstract will be reviewed by the conference committee and, if it is accepted, you will be able to present your work either as an oral presentation (Chapter 13) or as a poster presentation (Chapter 14).

All accepted conference abstracts are published as part of an abstract book and supplied to all conference participants. The participants use the abstracts to decide which oral presentations they would like to listen to and which posters they would like to see at the poster viewing session. Some abstracts are also published after the conference as conference proceedings in a peer-reviewed journal, and therefore count as a form of publication. Conference abstracts share the same features as the research paper abstract. That is, the abstract stands alone and includes the aims, methods, results, and conclusions of the work arranged sequentially.

10.4 Writing the abstract

Before writing an abstract you must familiarize yourself with the instructions supplied to prospective authors. These instructions will stipulate the word limit for the abstract, how it should be structured, and the format in which it should be submitted. When writing an abstract, comply with the guidelines provided **exactly.** If you do not, the abstract will be either rejected or returned to you for rewriting. Remember that readers will make a judgement on whether they should read your article in full (if it is a research paper) or view your oral or poster presentation (if at a conference) solely on the basis of your abstract. Therefore, make sure the abstract is written in a way that makes it easy to read and motivates the reader to find out more about your research.

Abstracts—some tips

- If you are writing an abstract as part of an article, then write the abstract last, after the rest of the article is complete, so that it truly reflects the content of the full article.
- Avoid lengthy background information and excessive descriptions of the methodology. The main focus of the abstract should be your results and conclusion.
- Structure your abstract so that it follows the sequence of aims of the study, methods, results, and conclusion. Do not confuse the arrangement, as this will make it difficult to follow your work.
- Do not cite references to other literature in the abstract, and avoid abbreviations as far as possible.
- Do not include illustrations or tables in the abstract. It should be a text-only description of your work.
- Conform strictly to the word limit and structure stipulated by any guidelines provided.
- Use the active voice.
- Use the checklist provide in Box 10.1 to review any abstracts you write.

BOX 10.1 Reviewing an abstract—checklist

Self-review any abstract you write using the following checklist:

- ❑ Does the abstract conform to the length limit stipulated by the guidelines?
- ❑ Does the abstract conform to the style (i.e. structured abstract or single paragraph) set by the guidelines?
- ❑ Does the abstract include the four elements: aims, methodology, results, and conclusions?
- ❑ Is the content of the abstract ordered in the prescribed sequence (i.e. aims, methodology, results, and conclusions)?
- ❑ Is the abstract free from references, illustrations, and unnecessary abbreviations?
- ❑ If you have used any abbreviations, have they been defined?
- ❑ Are there any parts of the abstract that can be edited to make it more concise (e.g. is all the information necessary, is there repetition of material, are there any long sentences that can be rewritten to make them more concise)?
- ❑ Is the abstract free from errors in spelling, grammar, and punctuation?
- ❑ Can the abstract be understood without reference to any further information?

You can improve your ability to write abstracts by working through the review examples in the Online Resource Centre.

Chapter 11

Preparing tables and figures

➲ Introduction

Your research is likely to generate a significant amount of data. These data need to be presented in a format that is meaningful and comprehensible to the reader. The aim of this chapter is to provide you with some general guidelines on presenting data in table and figure format.

Specifically, this chapter will:

- summarize the main features of figures and tables
- provide tips on preparing tables and figures
- provide some basic guidance on using software to generate and save tables and figures
- review the ethical reporting of data
- provide a checklist for reviewing the quality of tables and figures.

11.1 What are tables and figures?

Tables and figures are the two formats in which research data are presented (CSE, 2006):

- Tables are suitable for presenting either individual numbers or summaries of analysed data such as totals, percentages, means, and statistical analyses.
- Figures can include:
  - primary data such as photographs, micrographs, and electrophoretic gels and blots which show items. Items can include a whole range of data such as cells, atomic structures, DNA or protein bands, sequences, or spectra

- numerical data which have been transformed into graphs or charts, which show relationships between variables
- line drawings which can be used to illustrate concepts or models that are difficult to visualize.

Main features of tables and figures

- Each table and figure must present data **accurately and honestly.** This is discussed in Chapter 3, section 3.3.3 as part of ethical reporting of data, and summarized in section 11.4.
- Each table and figure must present data **clearly and simply** so that the data can be interpreted efficiently. For example, do not use 3D effects or background tints in line graphs and do not add shading to table columns or headings. To find out more about how to present quantitative data in a simple and clear format, see Tufte (2001).
- Each table and figure must be comprehensive so that it can be understood on its own without the reader having to refer to the main text of the article. This means that each table must include a title and explanatory footnotes and each figure must include a descriptive legend (see following points).
- Tables and figures are numbered consecutively as they appear in the text but in separate sequences. For example the first table is designated Table 1, the second as Table 2, and so on. The first figure is designated Figure 1, the second as Figure 2, and so on.
- Each table is given a descriptive title which is placed at the top of the table (with the table number). Explanatory footnotes, such as definitions of symbols used in the table or statistical significance levels, are placed at the foot of the table. Example 11.1 gives an annotated example of how a table is structured.
- Each figure contains a descriptive legend which is placed underneath the figure and includes:
  - the figure number
  - a title that is brief and concise but descriptive of the content of the illustration
  - a very brief description of how the data were collected and analysed
  - definitions of any symbols used in the figure
  - (and in some cases) key trends observed in the data, in particular any statistical differences. Annotated examples of how figures are structured are presented in section 11.2.2.
- Figures are commonly presented as composites made up of more than one illustration. In composite figures each individual figure is designated by a consecutive letter (a, b, c, etc.), and a single legend which includes information relating to each part is placed underneath the composite (section 11.2.3).

11.2 **Preparing tables and figures**

Before you start to construct your tables and figures, consider carefully the following questions:

1. **Which data should you present?** You do not need to report all the data you have generated during your study but only those which are relevant to the questions you have posed in the Introduction. Therefore, carefully select the data that most appropriately match your research questions. This does not mean that it is acceptable to omit data because they do not support your hypothesis. On the contrary, such data *must* be included. Only data that are not relevant to your research questions can be left out.
2. **What is your purpose for including a particular table or figure?** For each table or figure you include, there must be a clearly defined purpose for including it. The purpose may be:
   - to make comparisons between sets of data
   - to show trends and relationships
   - to show items as evidence
   - to illustrate difficult concepts and models which are difficult to visualize.

   In some instances, it may be possible to summarize your data in a few sentences without presenting it as a figure or table. If this can be done, then it is entirely appropriate to do so and then to refer to the data in the text as '(data not shown)'.
3. **If your data are numerical, are they suitable for presentation in graph or table format?** Some types of data sets lend themselves to presentation in tabular format; others are much better suited to presentation in graph format. Use tables if you are presenting individual numbers or summaries of analysed data such as means, and use graphs if you are showing trends or relationships between variables. *Do not duplicate the same data in both table and graph format.* Sometimes the journal editor's preference for tables or figures may be indicated in the instructions for authors, in which case you will have to comply with the format stipulated.
4. **If you use graphs to present your data, which type of graph should you use**? The four most common types of graphs for presenting data are line graphs, bar charts, histograms, and scatter plots. The type of graph you use will depend on your data set and is explained in section 11.2.2.
5. **How many tables and figures should you include?** This will depend on whether you are writing a paper for publication, a dissertation or thesis, or a poster or oral presentation. Some journals may stipulate the maximum number of figures and/or tables they will accept for publication, but others will leave it to your discretion. If

your journal of choice does not specify the maximum number of illustrations it will accept, look at some issues of the journal to get a rough idea of the number of tables/figures that are published per paper. In some journals, data that are useful but not essential for understanding your conclusions can be placed in the supplementary section of the paper (see Chapter 9, section 9.4.1).

For a poster or oral presentation, the number of tables and figures will depend on the space or time available to you. If you are writing up your results for a dissertation or thesis, then you will have space to include a larger number of illustrations. However, for each figure or table you include, there must be a purpose for including it. It is not appropriate to clutter your writing with non-essential figures or tables. If you include non-essential figures or tables and submit your work for assessment, you will be marked down. If you submit for publication, you will be asked to modify your paper or it may even be rejected.

6. **Are there figures that can be combined to produce a composite illustration**? Bringing figures together can be visually very powerful in highlighting key trends and comparisons as well as promoting a coherent and logical 'story' of the results. Therefore, consider carefully whether the understanding and coherence of your article could be improved if illustrations are grouped together.
7. **Which software packages should you use for analysing and graphing data**? There are a number of suitable software packages, which are described in section 11.3.

11.2.1 Preparing tables–Some tips

- Consider whether the data are suitable for presentation in tabular format. Remember, tables are used for presenting individual numbers or summaries of analysed data.
- Consider the size of the table; that is, how many columns and rows are required to present your data. The smaller the table, the easier it will be to compare the information presented.
- Keep the format of the table simple. If you are preparing tables for publication, your target journal will provide you with specific guidelines on how to format the table (e.g. not to use vertical rules). Follow these exactly.
- Arrange your data in the table logically. For example, give the control values first and then the experimental values.
- Be consistent in your use of decimal places throughout the table.
- Label each column heading in an informative way.

- Make sure any units of measurement are included either in the column heading (if the units of measurement for each value in the column are the same) or in the main body of the table (if they are different).
- Assign the table a number and write a brief descriptive title. The title and the number are placed at the top of the table.
- Add a brief footnote explaining any superscripts, symbols, or abbreviations you have used in the main body of the table. The footnote is placed underneath the table.

A good textbook to consult for assistance on how to construct tables is Nicol and Pexman (1999).

Example 11.1

A well-structured table

An annotated example of a well-designed table highlighting the basic features of a table is presented below (from Shi, Z., Zhang, H., Liu, Y., Xu, M., Dai, J. (2007). Alteration in gene expression and testosterone synthesis in the testes of male rats exposed to perfluorododecanoic acid. *Toxicol. Sci.*, **98**(1), 206–215, reprinted by permission of Oxford University Press).

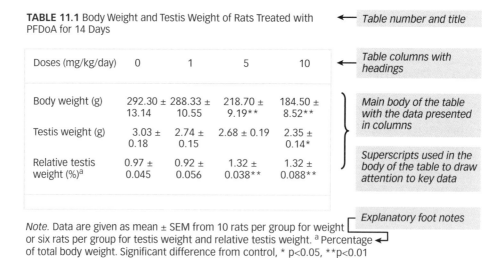

TABLE 11.1 Body Weight and Testis Weight of Rats Treated with PFDoA for 14 Days ← *Table number and title*

Doses (mg/kg/day)	0	1	5	10
Body weight (g)	292.30 ± 13.14	288.33 ± 10.55	218.70 ± 9.19**	184.50 ± 8.52**
Testis weight (g)	3.03 ± 0.18	2.74 ± 0.15	2.68 ± 0.19	2.35 ± 0.14*
Relative testis weight (%)[a]	0.97 ± 0.045	0.92 ± 0.056	1.32 ± 0.038**	1.32 ± 0.088**

← *Table columns with headings*

Main body of the table with the data presented in columns

Superscripts used in the body of the table to draw attention to key data

Explanatory foot notes

Note. Data are given as mean ± SEM from 10 rats per group for weight or six rats per group for testis weight and relative testis weight. [a] Percentage of total body weight. Significant difference from control, * $p<0.05$, ** $p<0.01$

This table is well designed because:

- It is numbered and includes a clear descriptive title which is placed at the top of the table.
- The data are presented in columns with clear column headings and units of measurement included.

- Explanatory notes including definition of the superscripts used in the main body of the table are placed at the foot of the table.
- The table is formatted and set out in a way that makes the information easy to follow.

11.2.2 **Preparing graphs**

There are different types of graphs that can be used for presenting data. The four most common types are line graphs, bar charts, histograms, and scatter plots.

Example 11.2

Line graph

A line graph is used to display a relationship between two continuous variables, the independent variable (which is plotted on the *x*-axis) and the dependent variable (which is plotted on the *y*-axis). In a line graph the individual data points are connected by a line to show how the dependent variable changes as the independent variable is changed. The figure (11.1) below has been annotated to show the basic features of a line graph.

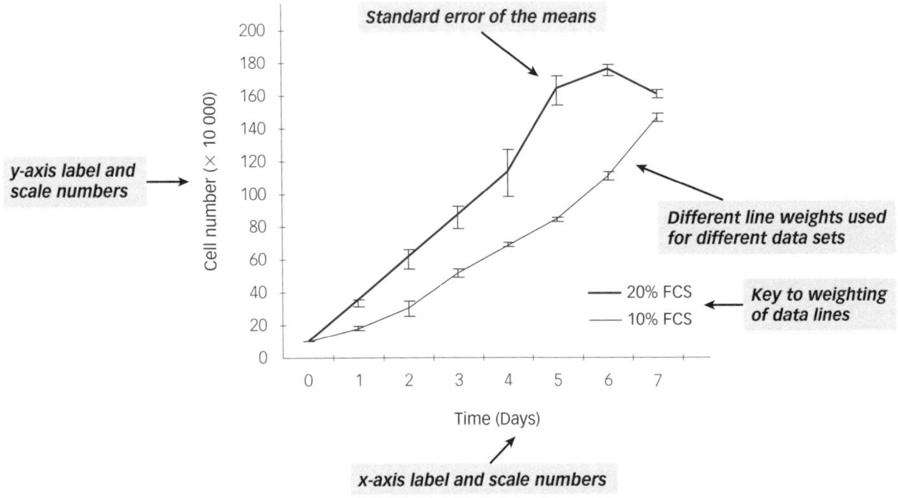

Figure 11.1 Effect of FCS (foetal calf serum) on cell growth. Y79 cells were supplemented with either 20% or 10% FCS in RPMI medium and incubated at 5% CO_2, 21% O_2 and 74% N_2 conditions. The cells were seeded at day 0 and the number of viable cells counted daily over a period of 7 days using the trypan blue exclusion method. Each point represents the mean of three cell counts. ← *Figure legend*

Example 11.3

Scatter plot

A scatter plot is also used to show a relationship between two continuous variables and is suitable for showing how individual data points are distributed within the *x–y* grid. The scatter graph can have a regression line added to show a trend in the data. It differs from a line graph in that the line may not connect each individual point but instead will be a line of best fit.

For an example of a scatter plot, see Figure 11.2:

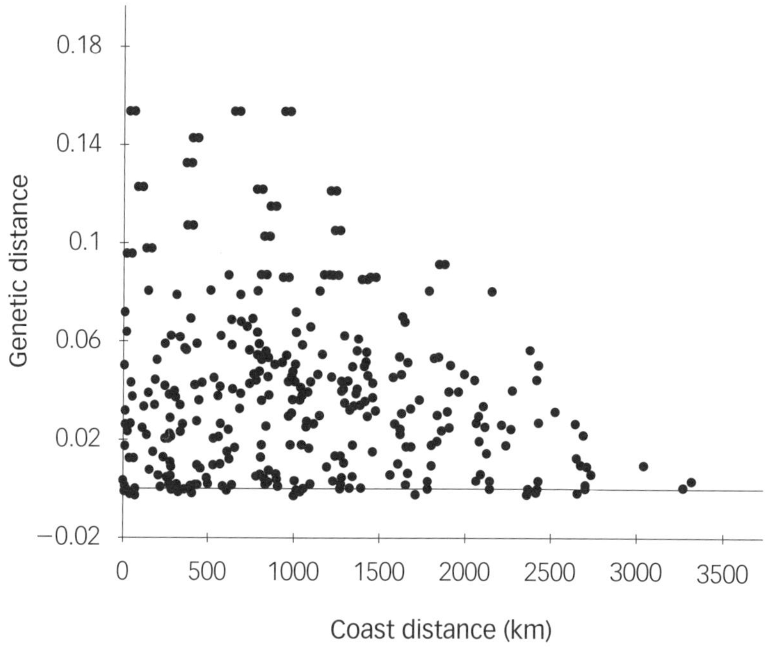

Figure 11.2 Genetic and coastal distances for 312 pairwise comparisons of sites on the British coast between St Abb's Head on the east coast, and Galloway on the west. There is a significant ($r = -0.159$, P 0.012) negative relationship (Mantel matrix test) between the distance matrices (unpublished data, Wilding and Grahame).

Example 11.4

Bar chart

A bar chart is used when at least one of the variables is a discrete variable. A discrete variable can be a category (group) such as species type, country, or gender, or it can be numerical such as a pulse count or population count.

For an example of a bar chart, see Figure 11.3:

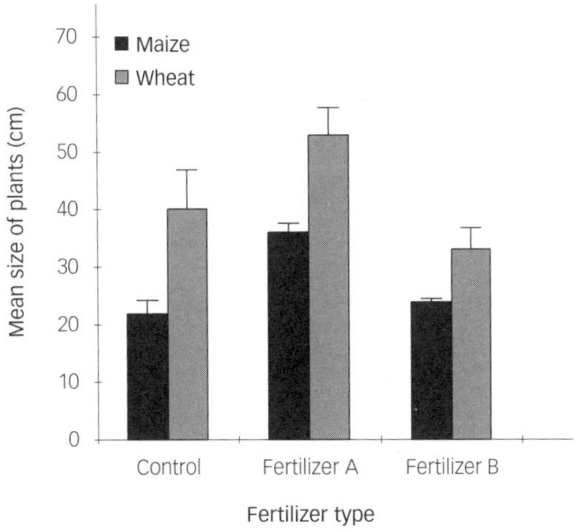

Figure 11.3 Effect of fertilizer type on the mean height of the maize and wheat plant. Maize (*Zea mays* L.) and wheat (*Triticum aestivum* L.) plants were sampled from individual 4 × 4 m plots treated with 100 kg/ha of fertilizer type (control (10 g/kg of N), A (20 g/kg of N), or B (30 g/kg of N)), 8 weeks after sowing. Each bar represents the mean height (± SEM) of 15 maize or wheat plants.

Example 11.5

Histogram

A **histogram** plots two continuous variables and represents a frequency distribution where each bar width represents a class interval and the area of the bar is the frequency with which data points within the class interval occur. The histogram is a good way of summarizing large data sets into a single graph.

For an example of a histogram, see Figure 11.4:

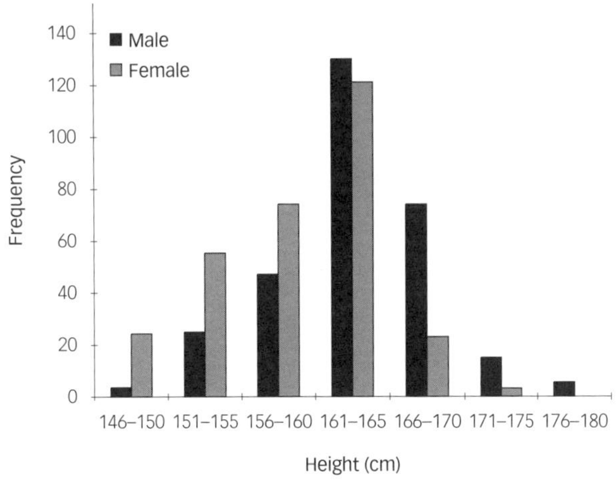

Figure 11.4 Height distribution of males and females aged 16. Males and females aged 16 were selected randomly from three geographical locations in the United Kingdom (London, Nottingham, and Edinburgh) and height (in cm) measured (N = 300 males and 300 females).

Useful guides to drawing different types of graphs can be found in the open access online resource produced by the Open University (2007) and the book by Nicol and Pexman (2003).

Preparing graphs—some tips

- Consider whether your data are suitable for presentation in graphic format. If yes, then select the correct type of graph.
- If you are using numerical scales for the *x*- and *y*-axes, make them appropriate in size and make sure the numbers increase in regular increments.
- Label each axis clearly and make sure that units of measurement are included where necessary. Labels on the *y*-axis are usually written vertically along the axis, as shown in the examples.
- Do not use more than four data sets on one line graph.
- When drawing line graphs with different data sets, use different line weights (e.g. solid, dashed, dotted) or different symbols (e.g. squares, circles, triangles) for each set of data. Do not use colour for figures that are intended for publication as it will increase the cost of publication. In any case, colour artwork is very likely to be reproduced in black and white, and you may lose essential information.
- Include a key explaining each line set (or bar shadings). This key may be placed in the body of the graph or adjacent to it.
- Include error bars for each data point (as necessary). If you have multiple lines on a graph and the error bars overlap, you may want to consider removing either the upper or lower half of each bar. This will make the graph easier to read and interpret.
- Include a descriptive figure legend. This is placed underneath the figure.
- If you are using several graphs that are related, then be consistent in the symbols, line weightings, or shadings that you use.
- Do not use background tints or 3D effects, and avoid the use of grid lines (unless it assists in reading the numerical values from the graph).

11.2.3 Preparing photographic images

Presenting electrophoretic gels or blots—some tips

- Add explanatory labels to each lane or column on the gel or blot.
- Label each significant band with the name and/or size.
- Make sure that positive and negative controls and molecular size markers are included on each gel or blot. If the markers or controls are not included (some published figures

may not show the controls or markers), then the original gel or blot with the markers and controls should be supplied in the supplementary information.

- You may crop a gel or blot to remove non-essential parts. If you do this, then you must make sure that all the important bands are retained and the original un-cropped gel or blot is supplied in the supplementary information.
- You may juxtapose vertical or horizontal sliced gels or blots which were not adjacent to each other in the experiment in order, for example, to make more effective comparisons. If you do this you must make sure that the boundaries between the separate gels or blots are clearly visible and the originals are retained.
- Include a descriptive figure legend. This is placed underneath the figure.
- For an example of a figure presenting an electrophoretic gel and a blot, see Figure 9.1.

Presenting micrographs—some tips

- You may crop micrographs to select the relevant information. However, you must retain the original un-cropped micrograph.
- Mark micrographs with arrows or labels to highlight significant components and to direct the attention of the reader.
- Add a scale bar to indicate the magnification on the micrograph (or you may place this information in the legend).
- Include a descriptive figure legend. This is placed underneath the figure.
- For an example of a figure presenting micrographs see Figure 11.5 (page 188).

Presenting maps of field sites—some tips

- Make sure you show the scale and orientation (e.g. indicate north) on the map.
- Make sure that the map you use has not been copyrighted, or that you have permission to use it.
- Mark the sites on the map from where the samples were taken.
- Include a descriptive figure legend. This is placed underneath the figure.
- For an example of a figure presenting a map, see Figure 11.6 (page 189).

11.2.4 **Preparing composite figures**

Individual figures are commonly grouped together and presented as a single composite figure. This is a visually powerful way of showing the strength of a particular conclusion and promoting the coherence of your article.

- When preparing composite figures, group figures together only if they are related to each other in terms of the research questions that they answer.

For full colour version, see Online Resource Centre

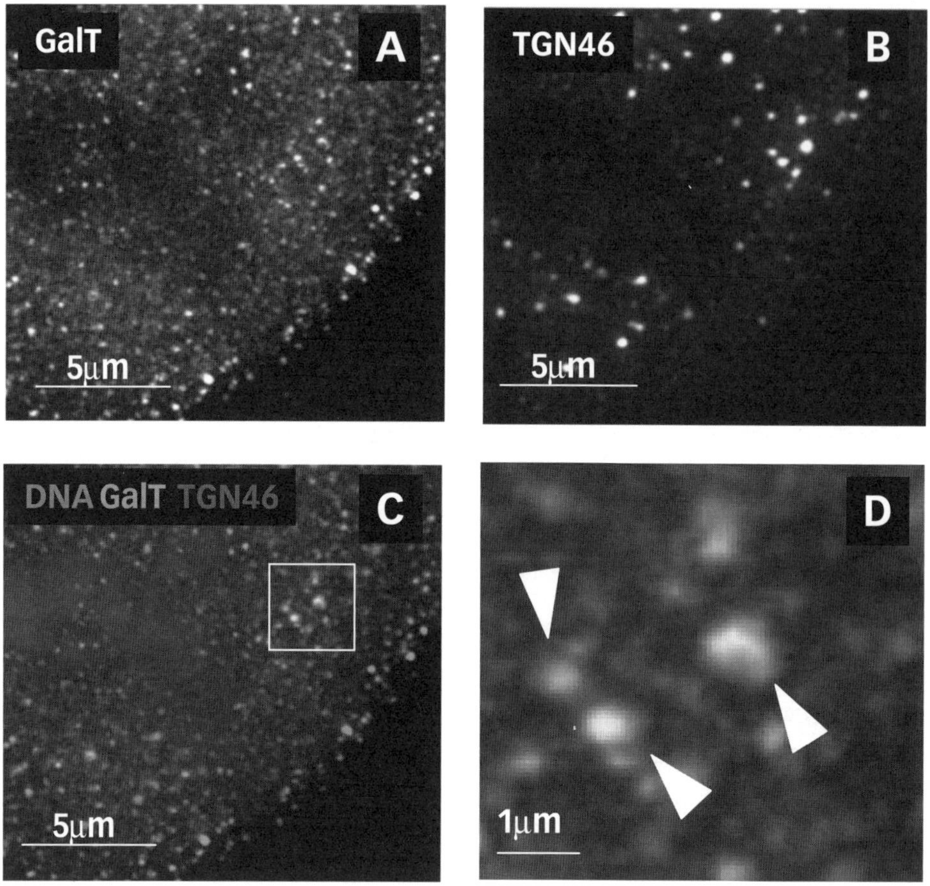

FIGURE 11.5 Localization of Golgi proteins during the cell cycle. Human epithelial cells were fixed and stained for the presence of Golgi-associated proteins GalT and TGN46 in order to determine whether different protein populations within organelles segregate as cells divide. Black and white images of GalT (panel A) and TGN46 (panel B) show these proteins exhibit a punctate distribution in dividing cells. Panel C shows a composite image where GalT (green) and TGN46 (red) appear to associate (boxed region, yellow pixels). Nuclear DNA appears in blue. The arrowheads in panel D highlight the structures within which GalT and TGN46 associate. Scale bars = 5 μm throughout, except panel D = 1 μm (unpublished data, Howell).

- Composite figures can consist of individual figures which show similar data types (such as multiple gel photos of different experiments) or figures which show a combination of data types (such as micrographs, photographs of electrophoretic gels, and graphs).
- Each individual figure is designated a consecutive letter (a, b, c, etc.) that is placed adjacent to the figure that it refers to.
- A single descriptive legend which includes information relating to each individual figure is placed underneath the composite figure.

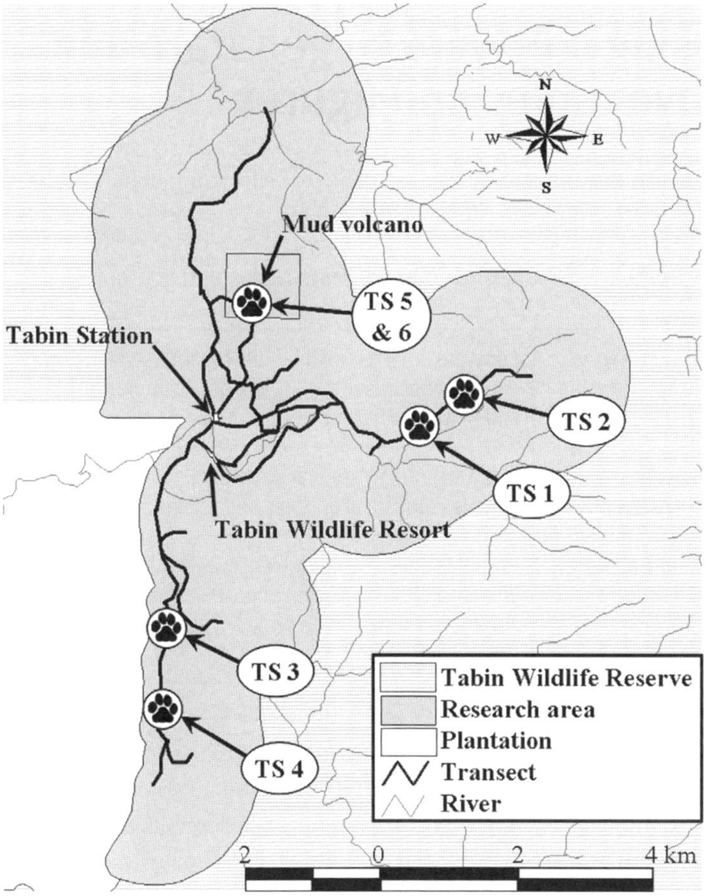

FIGURE 11.6 Location of the six track sets (TS) of clouded leopards (*Neofelis nebulosa*) within the Tabin Wildlife Reserve in the eastern part of the Malaysian state of Sabah in north-eastern Borneo (Wilting *et al.*, 2006) (reprinted by permission under the Creative Commons Attribution 2.0 licence).

Before constructing composite figures, look through examples of figures and tables published in papers. This will give you a good idea of how to present your data and may also give you some indication of things to avoid! An example of a composite figure which consists of multiple gel and blot photos was presented in Figure 9.1 and one consisting of multiple micrographs is shown in Figure 11.5. Although both these examples include groups of figures of similar data types, remember, it is also possible to compile multipart figures made up of a combination of data types (such as micrographs and graphs).

11.3 Using software to generate and save tables and figures

A number of software programs are available to help you produce high-quality tables and figures. Some examples are listed in Table 11.2.

TABLE 11.2 Software packages used to generate tables and figures

Data analysis and graph plotting software	Software such as Origin and Microsoft Excel are used for building data spreadsheets, calculating data, simple statistical analysis, and plotting different types of graphs
Statistics software	SPSS, SAS, and R are statistical packages used for general and complex statistical analysis of data and for graphing the data
Graphics editing software	Graphics editing software such as Adobe Photoshop, Paintshop Pro, and Corel PhotoPaint allow you to edit images. For line drawings use a program such as Adobe Illustrator or CorelDraw

If you are unfamiliar with the use of any of these software packages, or others that you may use during the course of your programme, then consult your university to see if they provide any training sessions. Many software manufacturers' sites also offer online tutorials or advice on how to use their software which you can browse through, or you can consult the software instruction manual.

Three basic aspects to consider when generating illustrations are file type, file size, and resolution.

File type

There are a number of different files types which are classified as either vector or bitmap files. Vector file formats are mainly suited to figures which contain line-based elements such as graphs, maps, and diagrams. Examples include Encapsulated Postscript (EPS), Adobe PDF, and the drawing features in Microsoft Office (Word, Excel, PowerPoint, Access). Bitmap file formats include PNG, TIFF, JPEG, and GIF. Bitmap files are better suited to photographs and micrographs.

File size

Image files, especially bitmap files such as photographs and micrographs, can be large and may have to be compressed to make the file size smaller. File formats use either lossless compression or lossy compression. In lossless compression the size of a file can be reduced without loss of quality, but in lossy compression, as the name suggests, compressing a file size results in loss of detail and hence the resulting image is of a poorer

TABLE 11.3 Image file formats

File type	File extension	Comments
EPS (Encapsulated Postscript)	.eps	Suitable for saving line drawings and images. Files saved in this format can be imported into some other programs (such as Adobe Illustrator) for modification (e.g. scaling and cropping). Files are usually saved in this format in their final form, when they are about to be incorporated into a printed publication
GIF (Graphic Interchange Format)	.gif	Suitable for saving simple images and line drawings only. Images saved in this format can be reduced in file size using LZW lossless compression without a reduction in quality
JPEG (Joint Photographic Experts Group)	.jpeg or .jpg	Suitable for saving photographic images in their final form after they have been modified (e.g. cropped, rotated). This is because a JPEG image is compressed each time a file is opened and saved, which results in loss of quality. JPEG format is therefore not suitable for saving an image which requires manipulation, but it is a good choice for saving a photographic image in its final form, for display on the web or for use in a slide presentation
PNG (Portable Network Graphics)	.png	Suitable for line drawings and photographic images. It works well in slide presentations. In addition, compression does not reduce the quality of the image
TIFF (Tagged Image File Format)	.tif or .tiff	Suitable for saving original images as well as manipulated images. The image can be manipulated, saved, and compressed (using LZW compression) without a reduction in the quality of the image. Suitable for photographs and line drawings, but may be very large files

quality. The different file formats and the type of compression used by each type is summarized in Table 11.3.

Always save your original image in a format which can be manipulated and compressed without loss of quality. TIFF is one of the most suitable file types for saving an original image. Only start manipulating an image after you have saved it and always retain a copy of the original, unmodified image. This is particularly important for two reasons: (1) you can return to it if you make a mistake during the processing stages, and (2) you can establish the authenticity of your image if your results are questioned (see Chapter 3 and section 11.4).

Image resolution

The resolution of an image is expressed as dots per inch (dpi). If you are preparing a figure for printing, then most journals will ask for a photographic image with a dpi of 300 at the final size. If the image is for the web, then a dpi of 72 or 96 at the final size may be sufficient. The resolution is affected by the size (that is, the dimensions) of the image. If the image size is increased, then the resolution of the image decreases and the quality of the image is reduced. Hence, a small image may have a resolution of 300 dpi initially, but if it needs to be enlarged during publication, the resolution will drop to below the 300 dpi threshold. You should ensure that any figure you produce will have the correct resolution at the final print size. This is best done by scanning your images or capturing your photos at high resolution (but remember this will create larger file sizes). If you submit a figure at a larger size, it will be reduced in size for publication. This will not result in loss of resolution but the legibility of the figure may be affected. You should make sure that the information on any figure you produce will still be legible at the final printed size if your figure is reduced during publication.

11.2.5 Submitting tables and figures

Journals will provide very specific information on how figures should be prepared and submitted:

- **File formats.** The most commonly accepted formats are TIFF, JPG, and EPS. If you have saved a figure or table in a format different from the type required by your target journal, then you will need to convert it to the required format.
- **Resolution.** Most journals will ask for a photographic image with a dpi of 300 and a line drawing at a dpi of 600 at the final size.
- **Whether the figures and tables should be embedded into the main manuscript or supplied separately**. For manuscripts intended for publication, figures and tables are usually supplied separately. If you are writing a thesis or dissertation, then the figures and tables are embedded in the main text.

Once you have compiled your tables and figures, review the relevance, accuracy, and quality of figures by working through the checklist in Box 11.1.

You can practise your ability to generate high quality figures by working through the review examples in the Online Resource Centre.

BOX 11.1 Reviewing data presented as tables or figures—checklist

- ❑ Have you chosen the best way in which to summarize your numerical data (e.g. averages, percentages, ratios)?
- ❑ Have you chosen the best format in which to present your numerical data (e.g. table or graph)?

- ❑ Are all the illustrations necessary for understanding your conclusions?
- ❑ Are there any illustrations that can be combined to produce a single composite figure?
- ❑ Does each table have a number, title, and explanatory footnotes (as necessary)?
- ❑ Does each figure have a number, title, and descriptive legend?
- ❑ Have you included standard deviations or standard errors where required?
- ❑ Are all figures and tables appropriately labelled (e.g. each column or axis) with the correct units of measurement where necessary?
- ❑ Have you explained any symbols, abbreviations, colour, and shading you may have used?
- ❑ Can each illustration be understood without reference to the main text of the article?
- ❑ If the figures are related, have you used symbols, shading, and line weights consistently?
- ❑ Have you saved the images in the correct format, resolution, colour mode, and size according to any instructions you have been supplied with?
- ❑ Will the figure and table information still be legible if it is reduced in size at publication?
- ❑ If you are reprinting a previously published figure or table, have you obtained permission from the copyright owner?

11.4 Ethical reporting of research data

When producing tables and figures you must take extreme care to ensure that you do not misrepresent your data in any way—for example, by excluding outliers in data sets, or by overexposing gels and thereby masking additional bands, or by using axis scales that distort how the trends are perceived. The latter is particularly important if you are using Microsoft Excel to produce graphs, as this software (originally intended for use in business, not in science!) has a tendency to generate axis scales that are not always appropriate for the data set. You must therefore always check the axis scales and adjust the proportions appropriately to produce an accurate account of your data.

When using graphics editing software you must keep the processing to a minimum. You should not in any way 'manipulate your image to enhance, clarify or conform to an expected result' (Nature Cell Biology, 2004; Nature Cell Biology, 2006). This is termed 'data beautification' and is a form of misrepresentation (see Chapter 3, section 3.3.3). The journal *Nature* (*Nature*, 2007) has produced some detailed guidelines on what is acceptable processing, which you should read along with the tips in section 11.2.3. These guidelines are relevant for producing figures for publication in any journal that you may wish to publish in, and for producing figures as part of assessments such as your dissertation or thesis. The article by Rossner and Yamada (2004) is also worth reading. This provides some specific examples of inappropriate practices based on real cases of digital manipulation discovered through inspection of digital images (micrographs and electrophoretic gels and blots) in a sample of papers submitted for publication in a journal. Remember, editors, assessors, and readers are not looking for pretty images or what you think the data should illustrate, but for real data that present your findings accurately.

References

CSE (2006). *Scientific style and format: the CSE manual for authors, editors, and publishers,* 7th edn. Council of Science Editors in cooperation with Rockefeller University Press, Reston, VA.

Nature. (2007). *Guide to publication policies of the Nature journals* [online]. Available at: http://www.nature.com/authors/gta.pdf (last accessed 7 Jan 2008).

Nature Cell Biology (2006). Editorial: Appreciating data: warts, wrinkles and all. *Nat. Cell Biol.,* **8**(3), 203.

Nature Cell Biology. (2004). Editorial: Gel slicing and dicing: a recipe for disaster. *Nat. Cell Biol.,* **6**(4), 275.

Nicol, A.A.M. and Pexman, P.M. (1999). *Displaying your findings: a practical guide for creating tables.* American Psychological Association, Washington, DC.

Nicol, A.A.M. and Pexman, P.M. (2003). *Displaying your findings: a practical guide for creating figures, posters and presentations.* American Psychological Association, Washington, DC.

Open University. (2007). *LDT_4 More working with charts, graphs and tables* [online]. Available at: http://www.open.ac.uk/openlearn/home.php (last accessed 7 Jan 2008).

Rossner, M. and Yamada, K.M. (2004). What's in a picture? The temptation of image manipulation. *J. Cell Biol.,* **166**(1), 11–15.

Tufte, E. (2001). *The visual display of quantitative information.* Graphics Press. Cheshire.

Wilting, A., Fischer, F., Abu Bakar, S., Linsenmair, K.E. (2006). Clouded leopards, the secretive top-carnivore of South-East Asian rainforests: their distribution, status and conservation needs in Sabah, Malaysia. *BMC Ecol.,* **6**(16), doi:10.1186/1472-6785-6-16.

Chapter 12

Writing a Master's dissertation or a PhD thesis

Introduction

The written report of the research work undertaken as part of a Master's programme of study is commonly referred to as a **dissertation** and the written report of the research undertaken as part of a PhD programme of study is commonly referred to as a **thesis**. Both the dissertation and thesis consist of a critical review of the background literature, a statement of the research objectives, the materials and methods used to test the research questions, the results obtained, and a critical discussion of the results. In the case of the PhD programme, an oral examination associated with the thesis commonly forms part of the assessment. Some Master's programmes also include an oral examination. The aim of this chapter is to guide you through the process of writing up your research findings in the form of a dissertation or a thesis (depending on your programme of study) and to provide guidance on the preparation of the oral examination associated with the research report.

Specifically, this chapter will:

- describe the basic requirements of a Master's dissertation and a PhD thesis
- list the components of the dissertation and thesis and review the expected content of each component
- describe a strategy for writing the dissertation or thesis
- provide guidance on preparing for the oral examination associated with the dissertation or thesis.

From this point onwards the term thesis is used to include both the Master's dissertation and the PhD thesis, to avoid cumbersome repetition. Where differences between the two exist, the distinction is made.

12.1 Requirements

The research undertaken as part of a (full-time) Master's programme in the UK typically lasts for about 3–12 months. Master's programmes may combine taught and research components or may be purely research-based (such as the Master's by Research). In a (full-time) PhD programme, the research work typically lasts for 3–4 years. In both programmes, the quality of the research work is assessed primarily on the basis of the content and analysis presented in the written report (and the associated oral examination where one is conducted). It is therefore extremely important that you are clear about the assessment outcomes you are expected to demonstrate in order to pass your programme of study.

12.1.1 Master's dissertation

As part of your Master's programme, you will conduct independent, in-depth research on a significant aspect (or different aspects) of your subject area and produce a dissertation on this work. The dissertation will be assessed according to your ability to:

- Formulate and present clear research questions which are well justified and informed by a comprehensive critical review of the literature relating to your area of study.
- Formulate a well-designed experimental strategy that rigorously tests the research questions posed. This includes defining the variable under study, the sample size, sample collection methods, and controls, and selecting and using appropriate methods of data collection and data analysis.
- Analyse the data obtained and present them clearly and accurately in both text and illustrative format.
- Critically interpret and evaluate the data obtained. This includes considering the different ways in which your data can be interpreted, discussing how your results compare with published studies, identifying limitations to your work, and making recommendations for further work.
- Produce a clearly written, precise, and coherent report that is accurately and completely referenced and free of spelling, grammatical, and punctuation errors.

12.1.2 PhD thesis

The PhD is the culmination of a much more extended period of study than the Master's programme and therefore includes more extensive and in-depth research on a significant aspect (or different aspects) of a particular subject area. It is a requirement of the PhD programme that the research will make a significant, original contribution which will advance the body of knowledge in the field of study in a meaningful way. It is normally expected that the PhD will lead to published work.

Your thesis will be assessed on your ability to:

- Formulate and present clear research questions which are well justified and informed by an extensive and in-depth critical review of the relevant literature in the field of study.
- Formulate a well-designed experimental strategy that rigorously tests the research questions posed. This includes defining the variable under study, the sample size, and sample collection methods and controls, as well as selecting and using appropriate methods of data collection and data analysis.
- Analyse the data obtained and present them clearly and accurately in both text and illustrative format.
- Critically interpret and evaluate the data obtained. This includes considering fully the different ways in which your data can be interpreted, discussing how your results compare with published studies, identifying limitations to your work, and making recommendations for further work. Your discussion (and your thesis as a whole) should demonstrate a full awareness of the depth, breadth, and complexity of the subject area.
- Produce a clearly written, precise, and coherent report that is accurately and completely referenced and free of spelling, grammatical, and punctuation errors.

12.2 Structure and content of the thesis

Theses are commonly structured into chapters, consisting of the Introduction, Materials and Methods, Results, and Discussion, presented sequentially in this order. An alternative to this structure is the 'thesis by published papers' format which is described in section 12.3.

For a Master's dissertation, you may be able to write up your results in a single chapter. However, for a PhD thesis, it is likely that you will have multiple research chapters with each results chapter addressing particular but related research questions.

In addition to these four chapter types, additional components of the thesis include:

- title page
- list of contents
- list of abbreviations
- list of illustrative materials
- abstract
- acknowledgements
- reference list
- appendices.

The content of each of these components is outlined below.

12.2.1 Title page

The title page typically includes the following information:

- The title of the thesis.
- The name of the author.
- The degree for which it is submitted.
- The name of the university and the department(s) where the work was conducted.
- The month and year of submission.
- A statement to the effect that the work is the candidate's own and that appropriate credit has been given where reference has been made to the work of others. (If you are including in your thesis material from a paper you have published, then the statement should declare this.)
- A copyright statement that states the ownership of the text and the intellectual property and the conditions of use of the material.

The exact wording for the last two statements will be supplied by your institution. **You should therefore check and comply with the wording exactly**. Sometimes the last two statements are supplied on a page following the title page. Again, your instructions will explain this.

12.2.2 Abstract

The abstract is a concise summary of the thesis, and outlines the research aims, the main methods, the main results, and the main conclusions of the study. Detailed guidance on how to write an abstract is provided in Chapter 10.

12.2.3 Introduction

This chapter critically reviews the background literature to your research area, states the main objectives of your research, and describes the approach taken to address the research questions. Overall, the introductory chapter should set your work in context and convince the reader of the need to investigate the questions you have proposed. For guidance on writing a review of the literature, refer to Chapter 7.

The introductory chapter should demonstrate to the examiner that:

- You are familiar with the literature in your field and have understood it.
- You are able to make judgements about which literature is important to include (e.g. main researchers in the field and key papers).
- You are able to review the literature critically and integrate information from different studies with different perspectives to formulate an independent view of the material.
- You are aware of the gaps in knowledge in the area and have positioned your work to answer some of these questions.

12.2.4 Materials and methods

This chapter describes the materials and methods used to answer the questions stated in the introductory chapter. The materials and methods section is typically arranged into three parts and includes:

1. A description of the materials—that is, the biological samples, chemical reagents, and specialist equipment used in the study. This section should also include how the materials were prepared and stored. Details of any ethical or statutory approvals obtained to conduct the work should also be included.
2. A description of the methods used and how the data were analysed.
3. A list of the names and addresses of suppliers of chemicals, reagents, and equipment used.

Chapter 9, section 9.4.6 provides detailed guidance on writing the Materials and methods section for a research paper. Remember, however, that the level of experimental detail supplied in a thesis is greater than that supplied in a research paper. It is a useful exercise to go through the Materials and methods section in a research paper and in a thesis and note the differences in the level of detail between the two. You should also read examples of theses in your subject area. Both these activities will give you an indication of the level of detail expected when writing the materials and methods section of a thesis.

The materials and methods should demonstrate to the examiner that:

- Your experimental strategy is robust and tests the research questions posed in the introductory chapter.
- The experimental details are sufficiently detailed and concise to allow another scientist to repeat the work.
- Your methods of analysis are appropriate for the data obtained.

12.2.5 Results

This chapter describes the results obtained using a combination of text, figures, and tables. The figures and tables present the data while the text describes what the data are showing by *summarizing the key observations and the main trends.* The results section is a *description* of your observations and therefore you should leave the *detailed interpretation* of your findings for the discussion section that follows. However, when describing your results it will be necessary to include at least some interpretation of the results, so that the reader can see how one set of results leads on to the next.

As already mentioned, for a Master's dissertation you may be able to write up your results as a single chapter of results. However, for a PhD thesis, it is likely that you will have multiple results chapters each addressing particular but related research questions. In this case it is common to make each results chapter self-contained so that it starts with a brief introduction that contextualizes the specific research questions addressed, briefly describes the methods used, and then presents and discusses the results obtained.

When presenting data in graphic or tabular format, make sure each figure or table is clear, accurate, and comprehensive enough to be understood without reference to the main body of the text. This means each figure must include a descriptive legend and each table must include a title and any necessary explanatory footnotes. See Chapter 11 for detailed guidance.

Each figure and table should be embedded in the main body of the text (unless the instructions for candidates state otherwise), and descriptions in the text should concisely and clearly describe the key findings, make comparisons as appropriate, and highlight the main trends. The description of the results should not repeat excessively information already presented in the figures and tables, nor should it describe experimental procedures. However, it is appropriate to provide indications of the methods used to place the result being described in context, without repeating the methods in detail. You should present your results in sections and add brief, informative sub-headings to add clarity. For examples of how to write text-based descriptions of your figures and tables, see Chapter 9, section 9.4.7.

The results chapter(s) should demonstrate to the examiner that:

- You have obtained high-quality data which you have analysed accurately.
- You are able to present data in the form of well-structured, comprehensive figures and tables arranged in a logical sequence.
- You have described what the results show accurately and concisely.

12.2.6 **Discussion**

The discussion chapter brings together all the different strands of your data to provide a composite picture of how they can be interpreted and how the data fit in a wider context. To achieve this you will need to:

- Briefly summarize your main research findings.
- Provide explanations for the findings.
- Compare your results with the results of other published studies to see how they support or conflict with them.
- Consider the wider implications (theoretical and/or practical) of your work.
- Discuss the limitations of your work.
- Make recommendations for further work.

As you write your discussion:

- Address the questions you have posed in your introductory chapter so that the reader can follow the extent to which your research aims have been met.

- Be critical in your interpretation of the results. Consider all the alterative conclusions that can be inferred from your data.
- Compare your findings with existing primary published primary literature and discuss the extent to which they support or conflict with your work.
- Discuss any limitations your study may have. If it is difficult to reach a particular conclusion on the basis of the factual evidence available, then say so.
- Highlight how your work is 'new'. This may involve proposing models that build on (or represent an alternative to) an existing model, or identifying how the work can be used practically.
- Avoid common mistakes such as stating conclusions that are not supported by factual evidence and speculating excessively. Some speculation is, of course, acceptable—for example, when proposing models. However, always differentiate clearly between what is speculation and what is supported by factual evidence.
- Divide your discussion into sections with brief, informative sub-headings to add clarity.

Overall, your discussion should demonstrate to the examiner that:

- You are able to lead the reader systematically through your line of reasoning to show the most plausible conclusions that can be inferred from your findings.
- You are able to integrate the conclusions from your individual results chapters to provide a coherent discussion of your work.
- You are able to situate your findings in the context of the wider literature.
- You are able to identify how your work extends the existing body of knowledge in the field and assess the significance of this.
- You are able to identify and evaluate the limitations of your work and make recommendations for future work.

12.2.7 **References**

The reference chapter lists all the references cited in the main body of the report. You will be working with a large number of references and therefore it is essential that you use reference management software to compile and manage your reference list (see Chapter 1, section 1.2). Overall, your reference list should demonstrate that:

- You have used authoritative sources of information in your work.
- You have conducted an extensive review of the literature in your field and therefore the references are wide ranging.
- You have included the key papers and key researchers in your field of work.
- Your knowledge of the field is up to date and therefore the most current literature is also cited.
- Your compiled reference list is accurate and complete.

Table 12.1 Outline of the additional components of a thesis

Component	Description
Acknowledgements page	This page acknowledges the people who assisted you in your work: typically includes acknowledgement of academic, technical, financial, or personal assistance
Table of contents	Lists all the items included in the thesis, including the title of each chapter, the titles of subsections within chapters, and the page numbers
List of abbreviations	Provides a key to all the abbreviations used in the thesis
List of illustrative material	A list of all the illustrative material included in the thesis, including figure and table numbers and titles, with their respective page numbers
Appendices	Contain supplementary material that is necessary to the reader in understanding the study but would break up the flow of the report if it was included in the main body of the thesis. Items in the appendices can include, for example, computer printouts of protein and gene sequences, questionnaires, and maps. Appendix items should be grouped together appropriately, numbered (e.g. Appendix A1, A2, Appendix B1, B2, B3, etc.) and cited in the main body of the text at the point at which they are referred to
Supplementary material	Supplementary information such as film or computer programs can be supplied on CD with the thesis. However, the thesis should stand alone without the examiners having to refer to this supplementary material to understand and examine your work. If it is necessary for the examiners to view the material on the CD, then, depending on the regulations of your university, you may be asked to make a case for the examiners to view the material

12.2.8 Additional components

An outline of the content of the additional components of a thesis—acknowledgements, table of contents, abbreviations, list of illustrative material, appendices, and supplementary material—is given in Table 12.1.

12.3 Thesis by published papers format

An alternative to the thesis format described above is the thesis by published papers. In this alternative format, the thesis consists of published research papers (or papers accepted for publication or submitted for publication) (see Chapter 9).

The thesis by published papers format typically consists of:

- an introductory chapter
- the published research papers (or accepted or submitted publications)
- discussion chapter

In addition to these three main parts, the thesis also typically includes additional components:

- the title page
- a table of contents
- an abbreviations page
- a list of illustrative materials
- the abstract
- acknowledgements
- references
- appendices.

For a thesis in this format, the introductory chapter should state the aims of the research and provide a critical review of the literature in a way that links the different research papers together so that the thesis reads as a coherent and integrated whole. Similarly, the discussion chapter should bring together the main discussion points of the individual research papers.

Some universities may allow the option of presenting a thesis in either the traditional chapter format or the published papers format. Others may allow only one format or the other. You must consult your university at an early stage in your research to find out the format in which your thesis should be prepared.

12.4 Strategy for writing your thesis

Writing a thesis takes a substantial amount of time, and always longer than you anticipate, so do not leave your writing until the last few weeks or months of your research work. Instead, you should write up aspects of your work as you are conducting your research. In particular, write up your materials and methods as you progress through your research and construct figures and tables as the data are generated. By writing continuously throughout your research, you are more likely to produce a well-thought-out, coherent report with the detail and attention required to cover such a substantial amount of work. In addition, producing figures as you go along will help you to understand your data and therefore help you adapt your work (if necessary) as your research progresses.

To assist you in writing your thesis, an 11-step strategy is outlined below, starting from the planning stages through to writing and finally submitting your work.

12.4.1 Step 1: be fully aware of ethical issues relating to the reporting of data

Chapter 3 discussed issues relating to ethical reporting of data, which you should review before you start to write your thesis. Specifically, you should know how to present your data accurately, how to reuse material that you have already published, and how to use other people's work accurately in your own work.

12.4.2 Step 2: read the instructions supplied by your programme of study

Before you start to write, read the instructions supplied by your institution and programme of study on the style, length, and submission format of theses. Some general guidelines are presented below. However, exact details vary between institutions and you should consult and comply with the instructions supplied by your own institution when writing up your work.

- **In which format should the thesis be submitted**? You may be required to submit your thesis in either the traditional or the published papers format, or you may have the option of choosing between the two format types (see sections 12.2 and 12.3).
- **What is the maximum length of the thesis**? You will be told the maximum word limits for the entire thesis as well as the word limit for the abstract.
- **What sections must be included and in which order should these sections be arranged**? The sections and order of arrangement are typically the title page, acknowledgements page, abstract, table of contents, list of abbreviations, list of illustrative material, introduction, materials and methods, results chapters, discussion, references, and appendices. For the published papers format, the materials and methods and the results chapter are replaced by the published papers. But again, you must read the instructions supplied.
- **What text style and formatting should be used?** This will depend on your institution, but commonly theses are written in Times New Roman font, point size 12, with either double line or one-and-half line spacing.

It is a good idea to review examples of theses previously submitted in your subject area to familiarize yourself with the expected structure and style before you start to write. You supervisor will be able to supply you with examples of previous submissions and you should also be able to access theses awarded by your institution via your university library. An additional source is the Networked Digital Library of Theses and Dissertations (NDLTD) and the Australian Digital Thesis Program (ADT). Both databases hold full-text electronic theses which can be accessed online (see Table 5.1). The NDLTD provides access mainly to theses awarded by US institutions (although participation by other countries is growing) and the ADT provides access to postgraduate theses submitted at Australian universities.

12.4.3 Step 3: define the scope of your thesis

As the thesis is divided into multiple chapters, you will need to plan your writing carefully so that each component part, as well as the entire thesis, is coherent and logical. To achieve this, first define the scope of your thesis by writing brief summaries (approximately 50 words) in answer to the following questions (adapted from Brown, 1994):

- What research questions have I investigated?
- What are the results?
- What are the conclusions?
- What methods were used to obtain the results?
- What gap in knowledge did the work address?
- What is known at the end of the study that was not known prior to starting the study?
- Why was it important to study this?
- What is still unknown about this area?

12.4.4 Step 4: produce a plan of your thesis

Once the scope of the thesis is defined, move on to produce a plan outlining the content of each individual chapter. Start by listing the chapter titles, chapter subheadings, and a list of the figures and tables that will form part of each chapter. At the same time include some notes on the possible content of each chapter (Gosling and Noordam, 2007). A sample thesis outline is shown in Table 12.2.

When deciding on the content, think about the aim of each chapter. For example, if it is an introductory chapter, then the aim is to review the background literature to your research. If it is a results chapter, then the aim is to describe and interpret your data. Considering the aim of each chapter and using the planning techniques described in Chapter 1 (section 1.3) should help you to identify relevant material for inclusion in your thesis.

Once you have produced a tentative outline of each chapter (which should be no more than a few pages in length), work out the approximate number of pages each chapter will consist of. Remember, this is an estimate only and the length is likely to change once you start to write, but estimating at this stage will allow you to schedule deadlines for completing each chapter.

12.4.5 Step 5: show your initial plan to your supervisor

Once the initial plan is complete, show it to your supervisor and get some feedback on it. You may need to revise the outline in light of the feedback you receive. It is also likely that the order in which you present your results will change as you begin to write. Do not be afraid to make the changes if a more logical order evolves that is different from your initial proposed plan.

TABLE 12.2 An annotated example of a PhD thesis outline. This outline is based on part of the work submitted for a PhD by Divan (2000)

Annotation	Outline
Title of thesis. This thesis uses the traditional format and is structured to include the four chapter types: the introduction, the materials and methods, results, and discussion	**Title: The role of p53 in retinoblastoma tumour progression**
The overall aim of the study is described and explained in the context of the background literature. The overall aim of the study is broken down into a series of measurable objectives	**Aims of the study** Retinoblastoma is a childhood tumour caused by the inactivation of the *RB1* tumour suppressor gene. In cells where pRb function is lost, the p53-dependent apoptotic pathway is commonly activated and the pRb-deficient cell is eliminated. Therefore, dual loss of pRb and p53 is usually required for a tumour to form. However, in humans, retinoblastoma develops in the presence of wild-type p53 and a fully functional apoptotic pathway. Thus the aim of this study is to investigate how a pRb-deficient retinal cell escapes apoptosis in the presence of wild-type p53 and a fully functional apoptotic pathway. Specifically, the objectives of the study are: • Characterize p53 gene status in retinoblastoma tissue sections and in retinoblastoma cell lines Y79 and Weri-Rb1. • Identify the effect of hypoxia and growth factors on p53 expression. • Characterize the effect of p53 expression on apoptosis and on its downstream effector, p21.
The main introduction lists the sub-sections that will form the introductory chapter	**Chapter 1: Introduction (35–40 pages)** **Section 1:** an overview of the eukaryotic cell cycle and how a breakdown in cell cycle control leads to tumour development. **Section 2:** the retinoblastoma tumour suppressor gene and its product pRb: how RBI was identified, the structure of pRb, the roles of pRb in tumour development. **Section 3:** the p53 tumour suppressor protein: identification, structure, functions, regulation defects in the pathway and consequence of the defects on cellular outcome. **Section 4:** the anatomy of the eye and the retina and how retinoblastoma develops. **Section 5:** aims of the study (see introductory paragraph above).
The second chapter is the materials and methods chapter: the main methods used in the complete study are listed	**Chapter 2: Materials and methods (approximately 35 pages)** This chapter will be divided into three sections: materials used, methods used, and the names and addresses of suppliers. Main methods used are: **Histochemistry:** (pRb, p53, p21, and Ki-67) using 50 formalin-fixed paraffin-embedded retinoblastoma sections, 2 frozen retinoblastoma samples, and 2 retinoblastoma cell lines (Y79 and Weri-Rb1) **Analysis of p53 gene status:** extraction of DNA from retinoblastoma sections and cell lines, primer design, PCR, sequencing, and analysis of sequencing results. **Analysis of cell viability:** using trypan blue exclusion method and flow cytometry (propidium iodide (PI)). **Analysis of apoptosis and cell cycle progression:** using flow cytometry (PI and Annexin V staining), morphological analysis, and TUNEL. **Analysis of protein expression:** using western blots, flow cytometry, and histochemistry. **Statistical analysis of histological staining data:** non-parametric Kruskal–Wallis test and non-parametric Spearman rank test. **Knock-out of p53 expression,** using siRNA oligonucleotides.
	Chapter 3: The role of p53 in apoptosis – an *in vivo* study (30–35 pages) **Aim:** histologically retinoblastomas are characterized by areas where there are 'sleeves' of viable tumour surrounding blood vessels which are in turn surrounded by confluent areas of apoptosis. The aim of this study is to investigate the spatial distributions of apoptotic cells, proliferating cells, and cells expressing p53 and p21 within the 'sleeve' structure.

(continued)

Methods: histochemical analysis of 50 formalin-fixed paraffin-embedded retinoblastoma sections (p53, p21, Ki-67), quantification and statistical analysis of histochemistry data and p53 gene sequencing.
Results: tumour cells displaced 100–200 μm away from the blood vessel undergo apoptosis and these regions correlate with low oxygen. In poorly differentiated regions of the tumour, p53 expression increases from the proximal regions (adjacent to the blood vessel) to the distal regions (away from the blood vessel) of the 'sleeve' as does the proportion of apoptotic cells, while p21 expression is restricted primarily to cells adjacent to the blood vessel. In contrast, in well-differentiated areas of the 'sleeve', this pattern of p53 expression is reversed so that the proximal zone contains many positive cells but the distribution of apoptotic cells and p21 expression remains the same. p53 gene sequencing data show that the p53 is wild type.
Figure 1. Micrographs showing retinoblastoma histology (differentiated and undifferentiated sleeves) and apoptotic cell morphology
Figure 2. Line graphs showing the distance of the blood vessel from the periphery of the viable cells in solid tumours
Figure 3. Micrographs showing p53, p21, Ki-67 expression in differentiated and poorly differentiated regions of the tumour
Table 1. Summary of quantification and statistical analysis data showing the distribution of apoptotic cells, p53 and p21 expression in the proximal, middle, and distal zones in the 'sleeve' structure
Conclusions: human retinal cells continue to proliferate in the absence of pRb and in the presence of wild-type p53. We propose that in poorly differentiated tumours, p53 expression is induced in response to a decrease in growth factors and oxygen, which leads to cellular apoptosis. In well-differentiated tumours, the apoptosis may be p53-independent. We also propose that p21 does not play a role in apoptosis but cell survival.

This thesis contains three results chapters. Each result chapter addresses a particular but related research question. Each is therefore structured to include the aim, the methods, the results, and the conclusions relevant to the chapter aims. The figures and tables which form part of each results chapter are also listed. When outlining each of the results chapters (as well as the introductory and final discussion chapter), use your responses to the questions in section 12.4.3

Chapter 4: The role of p53 in apoptosis—an *in vitro* study (30–35 pages)
Aim: to construct an *in vitro* model to explain the histological distribution of p53, p21, and apoptosis in the proximal and distal regions of the undifferentiated sleeve of the retinoblastoma.
Methods: grow retinoblastoma cell lines Y79 and Weri-Rb1 under different oxygen and growth medium concentrations and assess cell proliferation (using trypan blue exclusion method and PI staining), cell death (using Annexin V and TUNEL), and p53 and p21 expression (using flow cytometry and western blotting).
Results: cell viability decreases with decreasing oxygen and decreasing growth factor concentration, and p53 levels increase with decrease in oxygen and decrease in growth factors. p21 is associated with serum-rich medium but unaffected by oxygen levels. p53 correlates with apoptosis but not p21 expression.
Figure 1. Line graphs showing the effect of growth serum and oxygen concentration on cell growth
Figure 2. Line graphs showing the effect of growth serum and oxygen concentration on the proportion of apoptotic cells
Figure 3. Dot plot displays of Annexin V staining and staining patterns of TUNEL
Figure 4. Western blots of p53 and p21 expression
Figure 5. Relationship between p53 and p21 expression (line graph)
Figure 6. Relationship between p53 expression and the proportion of apoptotic cells (line graph)
Conclusions: the results support our *in vitro* observations that p53 levels rise as oxygen and growth factor concentrations are depleted, and this correlates with a corresponding increase in the number of apoptotic cells. The results also support the observation that p21 is not associated with the p53 apoptotic function.

(continued)

TABLE 12.2 Cont'd

Chapter 5: Effect of p53 knock-out on apoptosis (30 pages) **Aim:** to provide conclusive evidence that p53 modulates apoptosis in retinoblastomas, p53 expression was knocked out in Y79 and Weri-Rb1 cell lines and the level of apoptosis, p53 and p21 expression assessed. **Methods:** knock-out p53 using siRNA oligonucleotides directed at p53 and measure cell viability (PI), cell death (Annexin V and TUNEL), p53 and p21 levels (flow cytometry and western blots) at different serum and oxygen concentrations. **Results:** reduction in p53 levels correlates with decreased number of apoptotic cells and increased levels of proliferation (compared to control). p21 expression levels are similar in controls and p53 knock-outs. Figure 1. Western blots showing levels of p53 and p21 expression in response to p53 siRNA Figure 2. Proportion of p53, p21, apoptotic, and S-phase cells in siRNA-treated and untreated cells (when oxygen level is decreased) Figure 3. Proportion of p53, p21, apoptotic, and S-phase cells in siRNA-treated and untreated cells (when growth serum level is decreased) **Conclusions:** apoptosis observed in response to low oxygen and low FCS is mediated by p53 and therefore knock-out data support the observations made in Chapter 3 and Chapter 4.
Chapter 6 Discussion (approximately 20 pages) In poorly differentiated areas of the tumour, p53 expression is induced in response to a reduction in growth factor concentrations in combination with a low supply of oxygen, which triggers apoptosis independent of p21 activity. It is possible, therefore, that retinoblastoma cells may be rescued from apoptosis by survival factors delivered to the tumour by diffusion from the blood vessel. As retinoblastomas contain a fully functional pathway for limiting tumour formation, this pathway could potentially be manipulated for therapeutic purposes (as an alternative to existing surgical methods). Therefore, further work is required to characterize the downstream apoptotic pathway. We have also observed that in well-differentiated areas of the tumour p53 mediates cell cycle arrest through p21 transactivation, and apoptosis in this case may be independent of p53. This *in vitro* observation requires substantiating through *in vivo* studies
Chapter 7: References
Appendix: Sequencing chromatograms

The main conclusions of the thesis as a whole are summarized and suggestions for further work included

The final chapter lists all the references cited in the text

This thesis will include sequencing chromatograms which will be placed in the appendix at the end

12.4.6 Step 6: set deadlines for completing your report

Now that you have divided your work into smaller sections and agreed a plan with your supervisor, the next stage is to set deadlines for completing each section of the thesis. Setting deadlines is important as it will help you complete the writing up by the submission deadline and provide you with milestones for measuring how your writing is progressing. When setting deadlines, consider how much time you will spend daily on your writing. Writing a complete PhD thesis takes a substantial amount of time and always longer than you anticipate. You should, therefore, give yourself at least 6 months to write up your work. The amount of time you spend daily on your writing will depend on whether you are still conducting research or not. If you are, then you will need to balance your work (and other commitments) with your writing. In this case you could spend a few hours per day writing; for example, 1 hour in the morning and 1 hour in the evening. If your research is complete, then you will be able to devote more time to your writing. As you get closer to the submission deadline, it will be completely normal to write for 5 or 6 hours per day for 5 or 6 days a week.

Remember your supervisor(s) will be reading drafts of your work and providing you with feedback. Therefore, when setting deadlines take into account the time it will take your supervisor to return comments to you and then for you to consider the feedback and modify your drafts. Also take into account that you will self-review drafts of each chapter perhaps 4–5 times (if not more) before you reach a final version that is ready for submission, and build this into your time plan.

If you are writing up your research work as a Master's student, then aim to give yourself **at least** 3 weeks of **full-time writing** for a research project that lasted 3 months (and longer, if the project work was longer). Again, start writing before your research work is finished so that the report can be completed on time.

12.4.7 Step 7: set up a system for managing your files

You should already have a system for managing your files and references. If you do not, then set one up now: otherwise you could waste time trying to find things or lose important information and then waste time rewriting the lost material. Guidelines for managing electronic and print information were presented in Chapter 2 (section 2.3) and some specific points are summarized below:

- Make up back-up copies of your work regularly and keep copies in different locations (such as a storage device and on your home and work computers).
- Save successive drafts of your work, by dating or numbering versions, so you can keep track of the changes made and return to earlier versions if you need to.
- Consider writing and storing your files using a collaborative online tool such as Google Docs (see Chapter 15, section 15.4). This tool allows you to write and edit your work directly online, or you can upload edited documents from your computer to your online account. It is by default private but allows you to invite selected people to view or edit a document. This means you can grant editing rights to your supervisor(s) who can then access your files and make comments and corrections. Google Docs documents the changes made, when they were made, and by whom, and you can return to earlier versions of the documents. As the work is stored online, both you and your supervisor can access the files from any computer with an Internet connection.

12.4.8 Step 8: start writing your report

Your experimental notebook and the associated data (see Chapter 2, section 2.1) will form the basis of your thesis. If you have maintained a detailed record of your work, then the process of writing will be much easier.

As a PhD student, you will already have a substantial amount of relevant material which can be incorporated into your thesis. This material will include written literature reviews, progress reports, oral and poster presentations, and possibly published research papers. Therefore, consider carefully how this existing material can be integrated into chapters of your thesis.

Incorporating existing material into your thesis

You may not submit as the main content of your thesis any material which has been previously submitted for assessment either for the same degree or for another degree (either at the same institution or at a different institution). It is possible that a situation may arise where you are required to submit a piece of work on a topic you have previously submitted for **assessment purposes** (that is, a mark that contributes to the classification of a degree). For example, you may have already submitted an initial proposal of your research work which includes a substantial amount of literature review and then you write a thesis on the same project which also contains a review of the literature. In this case, it is advisable not to reuse the material in its original format but instead to transform it in a way that will make it an original piece of writing. You could do this by expanding a theoretical explanation or discussing a wider number of relevant studies in your work. Take advice from your tutors if you are unclear about the reuse of material you have previously authored (see Chapter 3, section 3.4).

If you have published a paper on your work, you will not be able to include a reprint of the published article to form a chapter of the thesis (unless you are writing your thesis in the published papers format). Instead, you will need to integrate the content of the paper so that it fits in with the research questions and the central arguments of your work. One way in which the content could be integrated is to place the materials and methods section of the paper in the methods chapter of the thesis, move parts of the introduction into the main introduction chapter and then extend on the background. You will of course need to reference the source appropriately and append a copy of the publication in the thesis.

The writing process

When writing your thesis, do not approach the writing in a linear manner—that is, starting with the title first and working through each component of the thesis sequentially to end with the discussion. If you do this, you will almost certainly lose sight of the central arguments of your work. Instead, start with the materials and methods section as this is straightforward and therefore easiest to write. Then construct the figures and tables and write the associated text-based descriptions. Follow this with the introductory and discussion chapters. Leave the writing of the abstract till the end. The abstract must truly reflect the content of the report and this is only possible after the introduction, materials and methods, results and the discussion components are complete. If you write the abstract at an earlier stage of the writing process, then go back and review it once the thesis is complete to make sure that it is truly reflective of the content. Once the abstract is written then compile the remaining parts of the report such as the appendices, acknowledgements, and table of contents.

There are different approaches that can be used to help you write the first draft of a particular chapter or section (Murray, 2006). One is to write down everything as it occurs to you. Another method is to list sub-headings and then add details below each sub-heading. Both methods will produce an initial, tentative outline of your work which can be restructured, fleshed out, and refined as your writing progresses.

As you write, conduct ongoing literature reviews so that the information included is the most current. Avoid the temptation to add a note such as 'insert reference' where actual references should be cited. If you do this, you may find that you can no longer identify the correct reference when you return to add the source. This can waste time unnecessarily. Use reference management software so that you can add the references in the text and build up your reference list as you write without it distracting your flow of thought.

The writing style

When writing a thesis, the same principles of accuracy, conciseness, clarity, and source referencing apply as for other forms of scientific writing (see Chapter 1, section 1.1).

- Report your work accurately. This includes using units of measurements, abbreviations, and scientific nomenclature (such as names of chemicals, genes, chromosomes, taxonomy) correctly and consistently. It also includes checking your figures and tables and any statements you make to see that there is no reason for anyone to misunderstand your work.
- Write concisely and clearly. Use short sentences and avoid unnecessary repetition. Try to use simple words as far as possible. There are, of course, times when technical words are necessary but, on the whole, avoid long and complex terms. Also avoid excessive use of bullet points and lists: your text should flow as a continuous piece of writing.
- Using an active voice when writing will make your thesis easier and more interesting to read. For example, it is acceptable to say '*I extracted protein from …*' instead of '*protein was extracted from ….*' and '*I conclude …*' instead of '*it was concluded …*'. When using an active style of writing, you may use the pronoun 'we' or the pronoun 'I'. The 'we' in this context will mean the single author; that is, you, and not a group of people. The choice between the two pronouns may depend on the preference of your university or department, and you should seek advice from your supervisor before you start to write. If your department recommends that you use a passive voice when writing your thesis, then you must of course do so.
- Make clear which is your work and which is the work of others by referencing your work accurately and completely. This includes citing the reference within the text at the point you refer to it and listing fully all the sources you have cited in the text at the end of your document in the form of a reference list (see Chapter 1, section 1.2 and Chapter 3, section 3.4).

Overcoming writer's block

Compiling a substantial piece of work such as a thesis takes time and effort, and there will inevitably be occasions when your writing stalls. When this happens, try the following techniques:

- Take a break from your writing. You can then return after a break refreshed and ready to continue.

- Work on a different section of your thesis. Working on different parts simultaneously can sometimes help clarify other parts of the thesis and facilitate writing generally.
- Talk to someone about your work. This can also assist in developing your ideas.

12.4.9 Step 9: review drafts of your work

You will need to self-review your chapters as you write them. Use the checklist provided in Chapter 9 (Box 9.1). This checklist is designed to review a research paper but applies equally to a thesis. It is also a good idea to review your draft against the assessment criteria the examiners will use to judge the quality of your thesis (described under the component parts in section 12.2) to see how sections of your thesis measures up. This will help you identify areas that require refining or adding to. Expect to self-review each chapter approximately 4–5 times (or more) before you reach a final version that is suitable for submission. It is a good idea to put the 'completed' sections away for 1–2 weeks before re-reading them. This fresh approach allows you to spot errors or grammatical inconsistencies, and if you do not understand parts then it is clear what you need to change.

In addition, your work will be improved by suggestions from your supervisor(s). When sending drafts of work to your supervisor(s):

- Make sure you ask them whether they would prefer a hard copy that they can write on directly or an electronic copy that they could mark up using the 'track changes' facility on word-processing software such as Microsoft Word. If you are sending an electronic copy by e-mail, then consider the size of your file (in particular if you have included images that take up a lot of memory space). Your supervisor(s) will not thank you for cluttering up their inbox! You can send large files by using FTP (see Chapter 2, section 2.2). The alternative is to reduce the size of files by saving images as compressed files (see Chapter 11, section 11.3) and by zipping multiple files together using software such as WinZip before sending them as an e-mail attachment. Of course, if you decide to use an online server such as Google Docs, then you will be able to store your larger files without the concern of taking up mailbox space.
- Make sure you correct your work for spelling, punctuation, and grammatical errors before sending to your supervisor(s) so that they do not waste time correcting these.
- Make sure you let your supervisor(s) know in advance when to expect drafts of your work so they can set aside the necessary time to go through the work. Make sure you also agree a time frame within which you can expect to receive the feedback from your supervisor(s). This will assist you in managing your writing schedule.
- Agree with your supervisor(s) how many drafts they will read and whether you will send drafts of individual chapters as you complete them or sections of your thesis (for example, one-third).
- Make sure you send drafts of your work well in advance of your submission deadline so that your supervisor(s) has time to read the drafts and you have the time to incorporate their suggestions into your work.
- Make sure you are aware of any time periods when your supervisor(s) will be away so that you can schedule this into your writing plan.

Once you receive the feedback, consider the comments carefully. You may decide to incorporate all of the suggestions into your report, or you may decide to include some but not others. If you decide not to include the changes suggested, then make sure you have justifiable reason for not doing so. If you require clarification or additional advice on any of the comments, then contact your supervisor(s) and discuss the received comments with them in a face-to-face meeting or via e-mail or telephone. If there are unexplained delays in receiving feedback, then follow up with your supervisor(s) to find out the reason for the delay. If the delay persists, then you should be able to obtain advice from the support structure that your department has in place for supporting students. This could involve talking to an advisor (or other identified person) allocated to mentor you during your programme of study.

12.4.10 Step 10: produce the final version

Once the chapters have been reviewed and redrafted, go through the entire document and check that:

- The pages are numbered sequentially.
- Formatting and text style is consistent throughout the document.
- All the figures and tables are cited in the text.
- All the references are cited correctly and completely.
- All the required pages are included.
- The dissertation is free from errors in spelling, grammar, and punctuation.

You are now ready to print your thesis, have the individual pages bound together, and then submit it.

12.4.11 Step 11: submit your thesis

You must submit your thesis on time and to the correct person. Make sure you submit the correct number of copies and that you complete and submit all the necessary forms (such as a declaration form to state that the work is your own) at the same time.

12.5 The viva

The *viva voce* examination, commonly abbreviated to viva, is an oral examination which follows the submission of the written thesis. It is an integral part of the examination process for a PhD student in many countries, and some Master's programmes may also expect students to attend a viva.

During the viva, the examiners will go through your work with you in a comprehensive manner and ask you questions relating to the background literature, your experimental strategy, the results obtained, and their interpretation. In addition to these specific areas, the examiners will also ask broader questions relating to how your results fit in with existing knowledge in the field, what is significant about your findings, and how your work could be developed further. The purpose of these questions is (Tinkler and Jackson, 2004):

- to assess whether you are able to discuss your work orally at a level which demonstrates that you have a deep understanding of your subject area
- to assess whether you have written the work you have submitted
- to clarify any areas that are weak or unclear in your report.

In the UK, the viva is typically conducted by at least two examiners, both of whom are specialists in your subject area. Of these examiners, at least one will be external to the institution where the research was conducted and one internal. Your supervisor is not normally present at the viva. In some countries, the viva format differs from the UK format described here and can consist of larger panels of assessors and sometimes even members of the public. The viva typically lasts around 3 hours for a PhD. In the case of a viva that forms part of a Master's programme, the length of time may be stipulated by your programme of study, but you can expect it to last at least 1 hour. During the viva you may be required to present your work orally to the assessors before the panel starts asking you questions, or there may be no formal presentation and hence the questioning will start immediately.

Examples of typical questions you can be asked at a viva are (adapted from Murray, 2003; Tinkler and Jackson, 2004):

- Why have you studied these particular research questions?
- Are there any limitations to the experimental strategy you have used?
- Why have you used this particular method of analysis?
- Is there any other way your data could be interpreted?
- What is original about your work?
- Which of your conclusions are the most significant and in what way?
- How could this research be developed further?
- What aspects of the work could be published?

12.5.1 Preparing for the viva

By preparing for your viva, you will feel much more confident and this will reduce the anxiety you may experience at the prospect of attending a viva examination.

Some guidelines to help you prepare

- Know the date and time of your viva and the place where the viva will be conducted.
- Know the format your viva will take. You may be expected to give a brief presentation of your work before the questions are asked, or there may be no formal presentation

and hence the questions will start immediately. If you are expected to present your work orally, then make sure you are well prepared to do so (see Chapter 13).

- Know who your examiners and what their roles are. In the UK, a viva is typically conducted by at least two examiners, the internal and the external examiners. Either the internal or external examiner may chair the viva or there may be an independent chair present. The rules vary from institution to institution, and you should read the regulations of your university so you know what to expect. Once you know who your examiners are, familiarize yourself with their research interests as this may give you some indication of the type of questions they could ask. This can be done relatively easily by conducting an author search on a database such as Web of Science.
- Read through your thesis and know the material well. As you re-read your work, anticipate the type of questions that could be asked and prepare responses to them. In addition, read the literature that has been published since your thesis was submitted so that you are completely up to date with the background to your research area.
- Go through a mock viva with your supervisor, or another academic member of staff who is familiar with your field of work. This will provide you with experience in discussing your work and allow you to see how a viva works in practice. Of course, the questions you are asked at the real viva will not be identical to those asked during the mock viva, but going through the process beforehand will at least give you the opportunity to discuss and justify your work.

12.5.2 During the viva

On the day of the viva, dress appropriately: most people wear formal clothes for a viva, and indeed at some universities academic dress may be prescribed. Take a copy of your thesis with you, as you will need to refer to it during the examination. Also take any additional notes you may have made during your viva preparation (e.g. potential questions that could be asked and the response to these). Give yourself sufficient time to arrive at the venue on time and try and stay calm. The relaxation techniques described in Chapter 13, section 13.5 should help you manage your nerves and combat any anxiety you may be feeling—and, remember, the examiners will aim to make the viva a positive experience for the candidate.

Some guidelines to help you during your viva

- Make eye contact with your examiners, sit up straight, smile, and speak clearly. You will project an air of confidence and this will make a positive impression on your examiners.
- Listen to the question being asked carefully. If you do not understand the question, ask for clarification.
- Wait for the examiner to finish asking the question before you start to answer.
- Answer questions succinctly and clearly, and keep your responses relevant.

- If you do not know the answer to a particular question, then say so. It is always better to admit you don't know something than to make up an answer.
- Expect to be asked questions that are critical of your work and challenge it. This is part of assessing how robust your understanding is. You should therefore not become defensive when such questions are asked, but instead try to provide a reasoned response.

12.5.3 On completing the viva

For a viva associated with a PhD (or Master's by research) programme, there are five possible outcomes that can follow the assessment of the thesis and viva. These are: pass without corrections, pass with minor revisions to the thesis, pass with major revisions to the thesis, referral (rewriting/interpreting data and sometimes further experimental work), or fail (you may in this case be recommended for a lower award).

You will be informed of the outcome either immediately after the viva or soon after. If you are asked to make corrections to your thesis (and most commonly students are), then make sure you know exactly what is expected of you in order to pass the programme, and when to submit the revisions by so that you are able to graduate on time.

For a viva associated with a taught Master's programme, your institutional regulations will explain how the viva contributes to the assessment of your programme and when the students will be informed of the outcome of the viva.

References

Brown, R. (1994). The big picture about managing writing. In: Zuber-Skerritt, O. and Ryan, Y. ed. *Quality in postgraduate education*. Kogan Page, London.

Divan, A. (2000). *p53, life, death and differentiation in retinoblastomas*. [PhD]. University of Sheffield, Sheffield.

Gosling, P. and Noordam, B. (2007). Mastering your PhD: writing your doctoral thesis with style. *Science*. Available at: http://sciencecareers.sciencemag.org/career_development/previous_issues/articles/2007_12_21/caredit_a0700183 (last accessed 28 May 2008).

Murray, R. (2006). *How to write a thesis*, 2nd edn. Open University Press, Maidenhead.

Murray, R. (2003). *How to survive your viva*. Open University Press, Maidenhead.

Tinkler, P. and Jackson, C. (2004). *The doctoral examination process: a handbook for students, examiners and supervisors*. Open University Press, Maidenhead.

Chapter 13

Delivering effective oral presentations

Introduction

As a scientist you will frequently be expected to communicate orally in different environments and to different types of audiences. It is therefore essential that you learn how to communicate your work in an informative and coherent way and deal effectively with any questions the audience may ask. The aim of this chapter is to guide you in the preparation and delivery of effective oral presentations.

Specifically, this chapter will focus on:

- how to plan an effective oral presentation
- how to organize and structure a talk
- how to design effective visual aids to support the presentation
- techniques for rehearsing the talk and managing your nerves
- how to use body language and voice to complement the content of the talk
- how to deal effectively with questions
- reviewing the success of your presentation.

13.1 The oral presentation—an overview

The main types of oral presentations you are likely to deliver are described in Table 13.1. They range from formal presentations to groups of experts, to informal presentations which may include less well-informed audiences. You may be speaking to individuals or to large groups of people. In all of these situations you must be able to communicate your information in an informative and coherent way so that the audience understands clearly what you have said. In addition, you must be able to deal effectively with any questions the audience may ask.

TABLE 13.1 Common types of oral presentation

Type of oral presentation	Brief description and purpose of the presentation
Presentation at a scientific conference	Scientific conferences, also known as scientific meetings, are professional meetings at which the most current research findings (not currently published as complete papers) are presented. These meetings are usually sponsored by subject-specific societies and attended by researchers working in the same field as the theme of the conference or in related fields. The meetings may be international, national, or local, and will consist of an expert audience with a knowledge base in the topic of the meeting. If you are presenting at a meeting then you will be expected to communicate your research findings in an allotted time, usually 10–30 minutes. These meetings are formal and are chaired by a chairperson, with questions and answers taken after each presentation
Departmental seminars	Most university departments or research institutes hold weekly seminars that cover a broad range of topics in the discipline. At the seminar, graduate research students and other invited speakers report their research findings orally. Each seminar is typically an hour long, which includes time for questions and answers at the end of the presentation. The audience consists of a range of people, some of whom may be working in the same field as the presenter and others who may be working in the same discipline but not necessarily in the same field. The purpose of attending the seminar is to broaden your knowledge of the discipline area you are working in and to place your work into a wider context. The purpose of presenting your work at seminars is to communicate your research to others
Journal clubs	These consist of small groups of people, usually members of a particular research group, who meet regularly to review papers published in the field they are working in. The format of these sessions may vary from club to club, but they typically involve one member of the research group orally presenting a summary of the paper and then leading a critical discussion of the work. The purpose of these sessions is to ensure that you are familiar and up to date with the current literature in your field of work
Presenting a case in support of or against an issue	You may be required to present a case in support of (or against) a particular issue. For example, if you are on a grant review panel, then you may need to speak about a particular grant proposal either in support of or against it receiving funding. If you are part of a debating group, then you may need to speak for or against a particular topic. The aim of this type of presentation is to present your argument to convince the audience of your point of view
Communicating science to the public	Not all the audiences you speak to will be scientific colleagues. On occasions you may be expected to speak to, for example, schoolchildren or adults who are members of the lay public and therefore largely unfamiliar with scientific information and terminology. If you are speaking to such an audience, then you will have to communicate simply with a minimum use of technical jargon so that the content of the talk can be understood by your audience

TABLE 13.1 Cont'd

Type of oral presentation	Brief description and purpose of the presentation
Presentation at a job interview	As part of a job interview you may be required to present a brief presentation on a particular topic. At these presentations, potential employers will be looking to see not only whether you have appropriate knowledge of the topic you are speaking on but also how you present yourself; that is, the personal impression you create by the way you speak, the way you stand, and the way you answer questions. The audience is likely to be mixed, consisting of specialists and non-specialists

There are six steps to making and delivering an effective oral presentation, and these are presented in Figure 13.1, starting with the planning stages of the presentation and ending with an evaluation of how you performed during the actual delivery of the presentation itself.

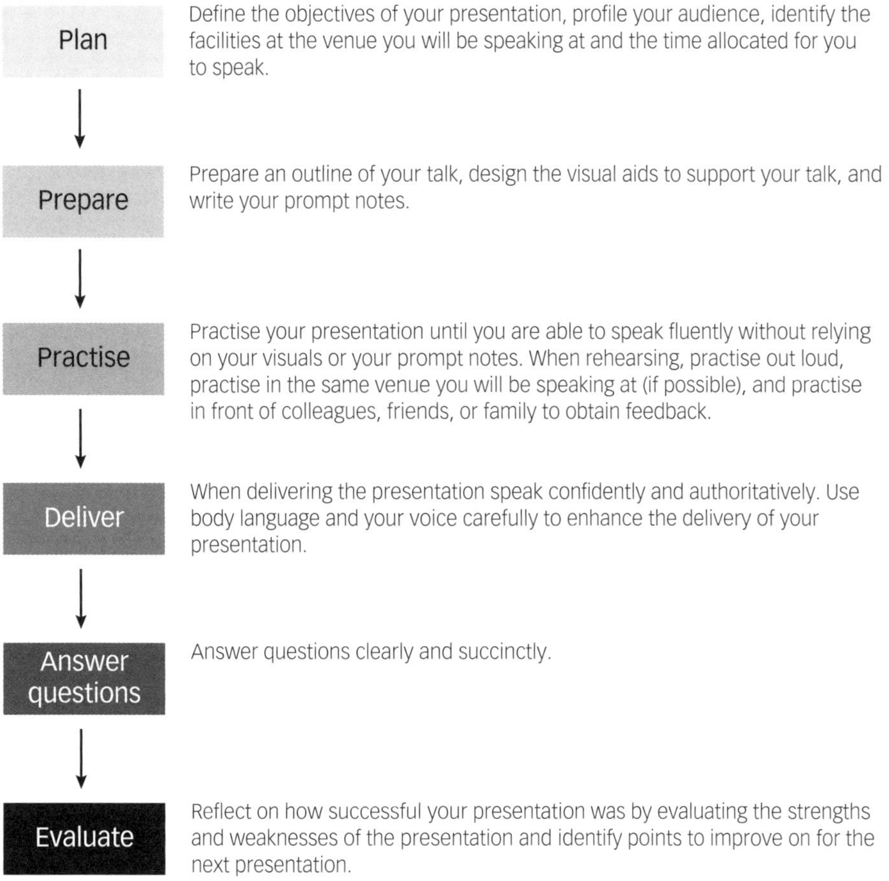
Plan
Define the objectives of your presentation, profile your audience, identify the facilities at the venue you will be speaking at and the time allocated for you to speak.
Prepare
Prepare an outline of your talk, design the visual aids to support your talk, and write your prompt notes.
Practise
Practise your presentation until you are able to speak fluently without relying on your visuals or your prompt notes. When rehearsing, practise out loud, practise in the same venue you will be speaking at (if possible), and practise in front of colleagues, friends, or family to obtain feedback.
Deliver
When delivering the presentation speak confidently and authoritatively. Use body language and your voice carefully to enhance the delivery of your presentation.
Answer questions
Answer questions clearly and succinctly.
Evaluate
Reflect on how successful your presentation was by evaluating the strengths and weaknesses of the presentation and identify points to improve on for the next presentation.

FIGURE 13.1 The six stages of an effective oral presentation.

The success of any presentation depends to a large extent on detailed planning, good preparation, and extensive rehearsing of the talk. Therefore, to deliver a good presentation, you will need to spend time on these pre-delivery stages. After the actual delivery of the presentation, you should evaluate how successful it was so that you can identify any areas you need to improve on the next time you are asked to speak. Each of the stages of delivering an effective oral presentation is described in turn below.

13.2 Step 1: planning your presentation

The first step to designing an effective oral presentation is to plan the presentation. When planning, address the following four questions:

1. **What** are the objectives of my presentation?
2. **Who** will I be speaking to?
3. **Where** will I be speaking and what facilities are available at the venue?
4. **How long** will I be speaking for?

13.2.1 What are the objectives of my presentation?

From the outset, be clear about the objectives of your talk; that is, what you expect to achieve. You will be able to define the objectives of your presentation by identifying the type of talk you are presenting and the audience you are presenting to. The common types of presentations you are likely to deliver as a scientist are listed in Table 13.1, together with a brief description and purpose of each type of presentation. Identifying the objectives at this stage is essential as it will help you focus on the content of the presentation and the format in which the talk could be delivered.

13.2.2 Who will I be speaking to?

You should try and find out as much as possible about the audience you will be speaking to, so that the content and the delivery of your presentation can be tailored in a way that is appropriate for the audience. If you do this, the audience is more likely to understand the material of your talk. The three key types of information you require about your audience are:

- **Who am I speaking to?** Depending on the type and purpose of the presentation, you may be speaking to adults, children, or a mixed audience.
- **What does the audience already know about the topic I am speaking on**? The knowledge base of the audience may be novice, expert, or mixed. By

identifying the knowledge base of the audience you will be able to pitch the material of your presentation at an appropriate level. For example, if you are speaking at a scientific meeting, most of the audience will be familiar with much of the background to your subject and therefore you will need to provide less background information to place your work in context. However, if you are speaking to a group of people who are not familiar with the topic area, then you will need to provide more background information so that your talk can be understood. It is not always easy to decide how much background information is necessary. One approach that you could use is:

- If you are presenting to an expert audience in your own field, then think about what you would have understood **before you started preparing the talk**. This is likely to be the baseline of the audience.
- If you are talking to a lay audience, then think about giving the talk at the level of a school student specializing in your broad subject area at an age of 15–16 years.

This should provide you with an idea of how much and what type of background information should be included in your talk.

- **How many people will be attending the presentation?** The size of the group you are speaking to will determine the format of the presentation. For example, if it is a large group, then the presentation is likely to be formal with less interaction between the audience and the speaker. If it is a smaller group, then there may be greater opportunity for a more interactive presentation.

13.2.3 Where will I be speaking, and what facilities are available at the venue?

You need to know what audiovisual facilities are available at the venue where you will be speaking and what the room size and layout is. This information will assist you in selecting the most appropriate delivery style and presentation tools (Table 13.2). If you are able to visit the room at this stage then you should do so. If not, ask the organizers of the presentation about the facilities and room size (or you may already have been supplied with this information—for example, by the conference organizers, if you are speaking at a conference). The number of people you will be speaking to is a good indication of what the room size and layout is likely to be. For example, if you are speaking at a formal conference with a large audience, then the venue is likely to be a tiered lecture theatre. If you are speaking to a smaller group then it is likely to be a flat room with people sitting around you in a semi-circle.

13.3.4 How long will I be speaking for?

Know exactly how long you are expected to speak for and **never** prepare a presentation that will go over your allotted time. Most presentations are followed by a

TABLE 13.2 Types of visual aids and the advantages and disadvantages of each type

Type of visual aid	Advantages	Disadvantages
Projection of computer-generated slides using presentation software such as PowerPoint	Can produce good-quality presentations that may include text and graphics Easy to add images, graphs, and tables from compatible software and digital cameras Easy to add, edit, and re-order slides Easy to prepare handouts of the presentations Ability to save the presentation on CD or memory stick which makes the presentation easily portable Suitable for both large and small audiences	Not particularly suitable for highly interactive audiences as there is limited scope for changing or adding slides during the presentation Colour schemes, slide designs, and animation have to be used with care as poor choices can be distracting to the audience Some of the features of the presentation may be lost if you present the slides at a venue which uses a different version of the software
Overhead projector (OHP) and transparencies	Easy to produce good-quality presentations by printing directly on to transparencies or by printing on to paper and then photocopying on to transparencies Ability to write on the transparencies by hand during the presentation Suitable for both small and large audiences	Transparencies are loose-leaf and can easily get out of order or fall to the floor
Slide projector and 35 mm slides	Can produce very good-quality visuals Suitable for both small and large audiences	Once produced, 35 mm slides cannot be modified, so can be costly to produce and update Slides need to be arranged and put in order in advance so that the sequence is correct and they are projected in the correct orientation Facilities for projecting 35 mm slides may not be available at many venues as they have been replaced to a large extent by computer-generated slides

TABLE 13.2 Cont'd

Type of visual aid	Advantages	Disadvantages
Flipchart/whiteboard	Suitable for small groups where there is high audience participation, such as brain-storming sessions Ability to write key points as you go along, which helps to maintain audience attention and makes the material easier to follow	Not suitable for large audiences If you spend too much time writing on the flipchart/board you may lose the interest of the audience You could end up with your back to the audience for much of the time
Video	Suitable for large and small audiences Can be a useful way to demonstrate examples of how things work, e.g. experiments	Requires time and skill to produce professional video clips
Models	Suitable for small audiences Very useful in demonstrating how things work, e.g. a model of the heart Very useful as a tool to engage the audience if the session is interactive	Not suitable for large audiences as it may be difficult to see from the back of a room Technical problems may arise, so you must ensure you are familiar with how the model works

question and answer session, so you need to know how much time has been allocated to the presentation itself and how much time has been allocated to the questions and answers. In addition, you should know what format the session will follow. For example:

- Is the session chaired? If it is, then you will be introduced to the audience by the chairperson who will also oversee the questions and answer session. If the session is not chaired, then you will be responsible for introducing yourself and managing the questions and answers.
- Will questions and answers be taken at the end or can questions be taken as you go along? If the session is formal, then the questions are generally taken at the end of the session. If the session is less formal or the audience is smaller then you may be able to take questions during the presentation.

Use the checklist in Box 13.1 to help with the planning of your presentation.

BOX 13.1 Planning your presentation—checklist

- ❑ What is the objective of my presentation?
- ❑ Who is my audience and what will they know about the topic I am speaking on?
- ❑ How many people will attend the presentation?
- ❑ What is their purpose in attending the presentation?
- ❑ How much time has been allocated to me to speak?
- ❑ What audiovisual facilities are available at the venue where I will be speaking?
- ❑ What is the size of the room and its layout?
- ❑ Will I need a microphone to be heard? Is one available at the venue?
- ❑ Is the presentation interactive or not?
- ❑ Will questions be taken at the end of the presentation or during the presentation?

13.3 Step 2: preparing your presentation

Once the planning stage is complete, you can move on to preparing the presentation. There are three parts to preparing a presentation:

1. Prepare an outline of your talk.
2. Prepare the visual aids you will use.
3. Prepare your notes.

13.3.1 preparing an outline of your talk

First prepare an outline of your talk taking into account the objectives of the presentation, who your audience is, and how much time you have allocated to you to speak. When planning your presentation, think carefully about the key messages you want to get across and build your presentation around these key messages. Use techniques such as brainstorming and mind mapping (see Chapter 1, section 1.3) to help you select the most relevant material for inclusion in your talk. If you are speaking at a conference, then you will have pre-submitted an abstract of your work and therefore you should use this abstract to plan the content of your talk.

Once the material is selected, then consider the sequence in which you will present that information, bearing in mind that oral presentations are commonly structured into three parts—the introduction, the main body, and the conclusion. This structure is sometimes described as '*first tell the audience what you are going to tell them, then tell them, and then*

TABLE 13.3 Features associated with the main parts of a presentation

Introduction	This is the opening of the talk; therefore tell the audience what the topic of the talk is and why it is important to speak on it. Then provide the relevant background to the topic so that the audience can place the information into a wider context. This part normally takes up approximately 10% of the talk time.
Main body	The main body covers the essential content of the topic you are speaking on. Depending on the topic area, this could include data, facts, theories, examples, arguments, analogies and anecdotes. This part normally takes up approximately 80% of the talk time.
Conclusion	The conclusion summarizes the key ideas discussed in the main body of the presentation. These are the main take-home messages and should link back to the introduction so the audience can see how the presentation comes together. This part normally takes up approximately 10% of the talk time.

end the presentation by telling them what you have told them.' The features associated with each part are listed in Table 13.3 (see also Walters and Walters, 2002).

An example outline of how a research presentation could be structured if you were speaking at a scientific meeting, for example, is shown in Table 13.4.

An example of a well-structured research talk in PowerPoint (with audio) can be downloaded from the companion website. The talk was delivered by a final-year PhD student, Claire McDonald at the Royal Entomological Society Postgraduate Forum, Rothamsted Research Institute (2008). The talk is entitled '*Herbivory in Antarctic fossil forests: modern day comparisons*'.

Beginning and ending your talk

When preparing your talk, think about how you will begin and end it. The audience will be at their most alert at the beginning of a talk, so you should use the opening to capture their interest by stating the topic and scope of your talk clearly. Similarly, you should conclude your talk in a way that reinforces the main points of your presentation and leaves the audience with your main take-home messages. Always plan a clear and decisive ending so that the audience is aware that you have finished. For example, signal to the audience that you are about to finish by using words such as 'to conclude' or 'to summarize'. You could then end the talk by thanking the audience for listening and/or by asking if there are any questions.

13.3.2 **Preparing your visual aids**

Once an outline is prepared then consider the visual aids you will use to support the delivery of your material. There are many different types of visual aids, and the main types are listed in Table 13.2 with a description of the advantages and disadvantages of each type.

TABLE 13.4 Structure of a research presentation for a scientific meeting

Title (author and affiliation)	Think of a brief and informative title for the presentation, and include your name and the name of the institution where you carried out the work when you pre-submit the abstract. This information will appear in the conference programme
Introductory background and problem studied	Identify clearly the objectives of your research and provide sufficient background to the study to place your work in context. Make sure you clearly establish which gap in knowledge you are addressing and why it is important to address it
Methods	Describe the experiments performed to investigate the problem. This part of the presentation should be brief and in a format that can be easily understood by the audience, e.g. in diagrammatic form
Results	The next few slides should present the key results of the research. As far as possible present the results as figures and tables so that relationships and trends can be clearly and easily visualized
Conclusions/summary	Add a short conclusion or summary of the work at the end to restate and reinforce the main points you have made
Further work	Pose further questions the research raises and which remain unanswered (optional)
Acknowledgements	Include a slide that identifies the people who funded the work and those who provided any additional support

When selecting visual aids for your presentation, take the following factors into account:

- **What is the purpose of using visual aids?** The purpose of using visuals is to support you in communicating your information to the audience. If used appropriately, they can be very effective in emphasizing key messages and illustrating concepts that are difficult to visualize. Do not use visuals to distract your audience with animation, or to avoid facing them. In addition, do not overload your slides with information and then just read aloud from the visuals.
- **What audiovisual facilities are available at the venue**? The facilities available at the venue may dictate your choice of visuals (see section 13.2).
- **What is the room size and how many people will you be speaking to**? Some visual aids are more suited to a smaller room and a smaller audience, while others are suitable for both small and large audiences and room sizes. Therefore, select audiovisual aids that are appropriate for your room and audience size.

Preparing slides

The key points to consider when preparing your visual aids are listed below. These refer specifically to the preparation of computer-generated slides using software such as Microsoft PowerPoint, but also apply to visuals such as transparencies and 35 mm slides. A selection of good and bad PowerPoint slides illustrating some of the points described below is shown in Figure 13.2. You may also want to consult the textbook by Fraser and Cave (2004) for additional advice on preparing computer-generated slides.

You can improve your ability to prepare high quality PowerPoint slides by working through the review examples in the Online Resource Centre

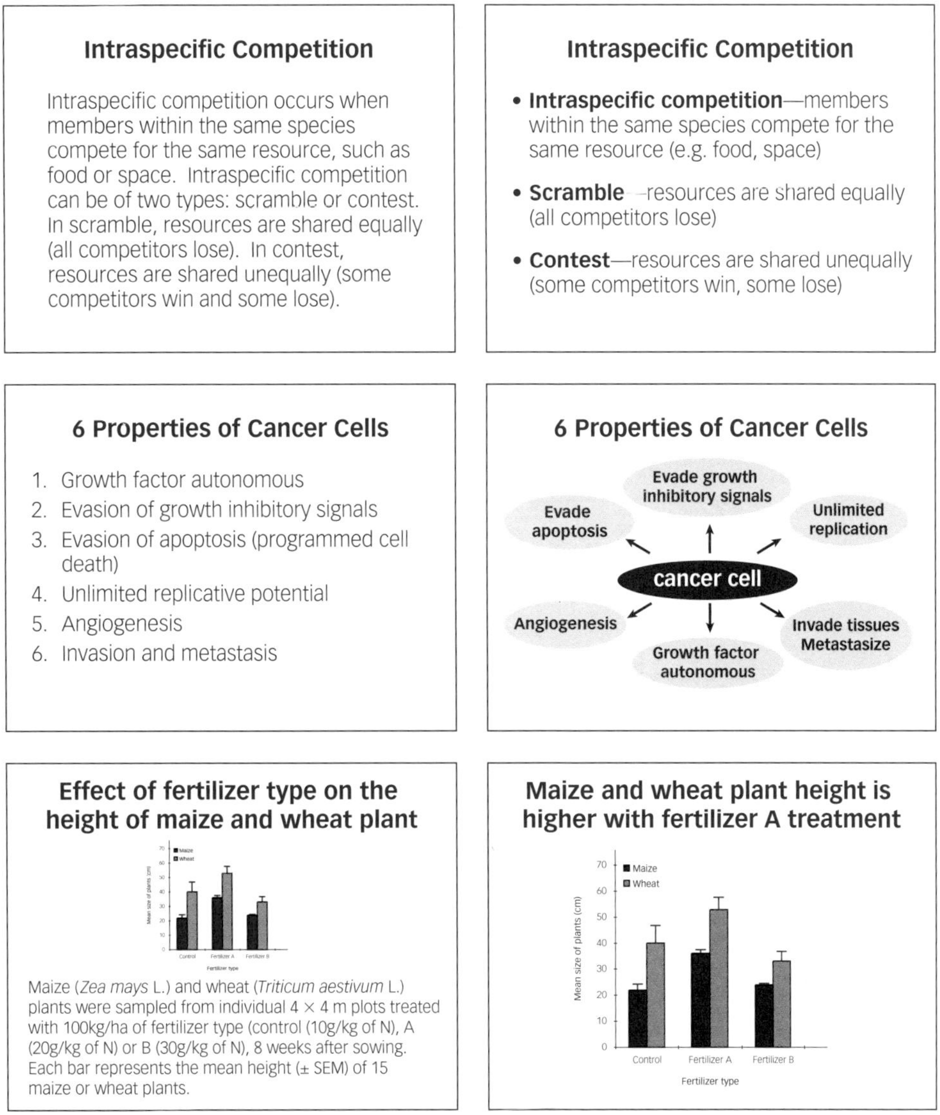

Intraspecific Competition
Intraspecific competition occurs when members within the same species compete for the same resource, such as food or space. Intraspecific competition can be of two types: scramble or contest. In scramble, resources are shared equally (all competitors lose). In contest, resources are shared unequally (some competitors win and some lose).
Intraspecific Competition
• Intraspecific competition—members within the same species compete for the same resource (e.g. food, space)
• Scramble—resources are shared equally (all competitors lose)
• Contest—resources are shared unequally (some competitors win, some lose)
6 Properties of Cancer Cells
1. Growth factor autonomous
2. Evasion of growth inhibitory signals
3. Evasion of apoptosis (programmed cell death)
4. Unlimited replicative potential
5. Angiogenesis
6. Invasion and metastasis
6 Properties of Cancer Cells
Evade growth inhibitory signals
Evade apoptosis
Unlimited replication
cancer cell
Angiogenesis
Invade tissues Metastasize
Growth factor autonomous
Effect of fertilizer type on the height of maize and wheat plant
Maize (Zea mays L.) and wheat (Triticum aestivum L.) plants were sampled from individual 4 × 4 m plots treated with 100kg/ha of fertilizer type (control (10g/kg of N), A (20g/kg of N) or B (30g/kg of N), 8 weeks after sowing. Each bar represents the mean height (± SEM) of 15 maize or wheat plants.
Maize and wheat plant height is higher with fertilizer A treatment
Maize
Wheat
Mean size of plants (cm)
0
10
20
30
40
50
60
70
Control
Fertilizer A
Fertilizer B
Fertilizer type

FIGURE 13.2 The slides on the left are poor. The slides on the right show better ways of presenting the same information.

Limit the amount of information per slide

Think carefully about the purpose of each slide and aim for one central message per slide. Do not overload your slides with too much text or too many graphics, or both. Always aim for simplicity and clarity when designing your slides.

Aim for a combination of text and illustrations in your presentation

Consider whether text is suitable for the information you want to convey, or whether an illustration would convey the message more convincingly. For an effective presentation, aim for a combination of text and illustrations (e.g. tables, graphs, line drawings, and photographs). PowerPoint allows you to import images, tables, and figures from a wide range of other software packages (such as Microsoft Excel, Origin, or Adobe Photoshop) which makes the compilation of your slides much easier. For guidance on preparing accurate and comprehensive figures and tables of your research findings, see Chapter 11. However, bear in mind that figures and tables prepared for display on slides must be much simpler than those prepared for publication in papers so that the trends or items can be visualized quickly and easily.

Preparing suitable tables and figures—some tips

Tables

- Keep the table simple and the values to a minimum. You could highlight individual or sets of data values to direct the attention of the reader to particular trends.
- Label each table column with an informative heading and include units of measurement (as necessary).

Graphs

- In a line graph, use no more than three lines of data on a single graph.
- In a line graph, use bold lines and contrasting colours for each data line. Avoid using dotted lines and pale colours as these are difficult to read.
- Use the same colours for the same treatments in different graphs.
- Label each axis, include units of measurement, and add standard errors (if necessary).
- Include a key to explain the different data sets.

Micrographs, maps, electrophoretic gels, and blots

- Crop the image to select only the relevant information.
- In a gel or blot, add explanatory labels to each lane.
- In a gel or blot, label each significant band with arrows, name, and/or size (as appropriate).
- Mark micrographs and maps with arrows to highlight any significant features to direct the attention of the viewer.

Use bullet points when writing text-based slides

When writing text-based slides use bullet-points rather than using paragraphs of continuous text. Aim for a maximum of six lines of bullet points per slide.

Use informative titles

Make sure that each slide has an informative title so that the listener can easily identify the content of the slide they are viewing. If it is a data slide, the title will often summarize the findings or outcome of the data presented in that slide.

Use fonts that are simple and easy to read

Sans serif fonts, such as Arial, Tahoma, or Verdana are the most legible fonts for presentations. Avoid fonts which imitate handwriting such as Handstroke or Bradley Hand ITC, and decorative fonts such as Snap ITC or Ramona. These are extremely difficult to read when projected on to a screen. However, the handwriting font Comic Sans can be read relatively easily and is therefore acceptable. If you are unsure about using a font, try it out: project it on to a screen and then assess how easy it is to read. Once you have selected a font, use it consistently throughout your presentation.

Use a type size that is large enough to be read at the back of the room

It is best to use a size of at least 36pt (usually 40–44pt) for titles, and at least 24pt for the main text.

Be careful with the use of colours

When using colours, use them purposefully; for example, you could use green for positive features and red for negative features (but remember that 10% of European men are red/green colour blind). When selecting colours, select those that harmonize with each other—red and green is a good combination of colours that work well together, as do orange and blue and green. Do not use colours randomly or without reason. Use strong colours, not pastels. If you need to show a lot of different types of line, use dotted or dashed lines in strong colours rather than pink and yellow, for example.

Use simple backgrounds and colour schemes

When selecting the text and background colours for your slides use a colour combination that will increase the readability of your slides. Dark backgrounds require light text colours and light backgrounds require dark text. If you are presenting in a darkened room, then aim for a dark background with light text; for example, a dark blue background and white text. If you are presenting in a well-lit room, then aim for dark text on a light background; for example, a pale blue background with black text. Use simple backgrounds and avoid multiple colour schemes.

Use a consistent template

When preparing your slides use a single template and stick to it. Avoid the temptation to change the colour scheme or background design with every slide or every few slides. If you have an institutional template then you could use that or, if you speak regularly, then you may want create your own template to use.

Avoid excessive animation, sound effects, and transitions

Avoid using animation and sound effects unless there is a purpose for using them and they would add to the understanding of your presentation. Do not use them merely for entertainment as they can be distracting to the listener who may lose your central message. Many people use transitions to add each bullet point on a slide, or between slides. This is a matter of personal preference, but again, if you use transitions, use a simple style and be consistent throughout the presentation.

13.3.3 Writing your prompt notes

Once you have prepared an outline for your talk and designed the visual aids, then think about what you will say about each slide. At the initial stage you may decide to write out your speech in full and then condense this into a series of prompt notes as you practise your presentation. Prompt notes can be written on index cards, which are easy to carry and hold when delivering your presentation (but don't drop them when you are on stage!). Alternatively, if you are using presentation software such as PowerPoint you can write your notes under each slide and print these out as one A4 sheet for each slide of your presentation. Irrespective of the format of your prompt notes, never read from your notes when delivering a presentation. If you do, the audience will lose confidence in you as a speaker. The role of the prompt notes is to *prompt* you in case you forget some essential information during the actual delivery of the presentation.

The checklist in Box 13.2 will help with preparing your presentation.

BOX 13.2 Preparing your presentation—checklist

Planning an outline

- ❑ What is the central message you want to convey to the audience?
- ❑ What material should you include to convey this message to the audience?
- ❑ In which sequence should you arrange the information?
- ❑ Which visuals will you use to support your presentation?
- ❑ Is the amount of information appropriate for the time you have been allocated?
- ❑ How will you introduce your presentation?
- ❑ How will you finish your presentation?

Reviewing the content and structure of your presentation

- ❑ Is the presentation structured with a clear introduction, main body, and conclusion?
- ❑ Is the material appropriate for the knowledge base of the audience?

- ❏ Is the sequence of information the most logical for an effective flow of ideas?
- ❏ Is the information (text and illustrations) accurate? Have you used facts to back up any assertions you have made?
- ❏ Are the objectives of the presentation clearly defined?
- ❏ Is the central message of the talk clear?
- ❏ Do you have a strong beginning and a strong ending?

Reviewing your slides

- ❏ Is every slide necessary or can some be deleted?
- ❏ Are any additional slides necessary to make your talk more comprehensible?
- ❏ Do the slides have informative titles so that the reader can identify the content of the slide easily?
- ❏ Is there a good balance of text and visuals throughout the presentation?
- ❏ Is there an appropriate amount of text/graphics on each slide without overloading?
- ❏ Will the audience be able to read your slides with the colour scheme you have used?
- ❏ Is the font clear enough to be easily read at the back of the room?
- ❏ Is the type size large enough to be easily read at the back of the room?
- ❏ If you have added animation to your presentation, is it purposeful and necessary?
- ❏ Are your slides free from errors in spelling, grammar, and punctuation?
- ❏ Have you checked your figures and tables for completeness?

  Are graph axes labelled?

  Do all table columns have headings?

  Are units of measurement included?

  Are standard deviations or standard errors included where required?

  Is there a key explaining any symbols, shadings, or line styles you have used?

  Are any bands and lanes clearly labelled?

  Are any significant features on photographs/micrographs clearly marked?

13.4 Step 3: practising your presentation

The key to delivering a successful presentation is to practise it. The benefits of rehearsing your presentation are:

- You will be able to see how your visuals work in practice. In particular, if you are using animations and transitions, you will know where they are so that you are not caught out during your presentation.

- You will be able to see whether your presentation fits into the time allocated to you to speak.
- You will become more familiar with your spoken words and this will make the delivery of your presentation smoother. This will also mean that you will be able to deliver your talk without reading from your prompt notes or from your slides.
- Your confidence level will increase and this will reduce the trepidation you may be experiencing at the prospect of speaking in public.

When rehearsing, use the following tips to help you refine your talk.

- **Practise out loud.** Always speak out loud when practising your presentation. As you practise, check that (1) your speech matches the visuals you are speaking about and (2) your presentation fits into the allotted time. At the initial stages of rehearsing, you may need to delete, rearrange, or add to your visuals and to your speech so that both are completely coordinated. You will also need to think about how you can move from one slide to the next smoothly. You could use, for example, transition terms such as 'now we can move on to discuss . . .' or 'now we have established . . .' Aim to rehearse your presentation until you are able to speak smoothly to your visuals without reading from the slides or from your prompt cards. This could mean practising your presentation a minimum of 10 times.
- **Practise in the room you will be speaking in**. If at all possible, rehearse in the venue that you will be speaking in, or in a room of a similar size with similar facilities. The advantage of speaking in the same room with the same facilities is that you will be able to decide in advance where you will stand in relation to the audience and familiarize yourself with the audiovisual set-up (e.g. how the microphone works, how the computer and projector system works, and where the lights are in case you need to dim the room).
- **Practise in front of friends and colleagues**. Initially you will practise in an empty room. However, you should also practise in front of colleagues or friends or family to get some feedback on your presentation. Use the evaluation checklist to review your presentation at this stage.

BOX 13.3 Evaluating your presentation—checklist

- ❑ Did I introduce my talk clearly so that the objectives of the talk were clear?
- ❑ Was the content of the talk pitched at a level appropriate for the audience?
- ❑ Was the content arranged in a logical manner so that it was easy to follow?
- ❑ Did I convey the central message of my talk clearly?
- ❑ Did I finish with a strong ending?
- ❑ Did I use visual aids in a way that supported and enhanced my presentation?

- ❑ Was my presentation smoothly delivered without reading from written notes?
- ❑ Did I stand upright without fidgeting?
- ❑ Was the presentation appropriately paced so that it was neither rushed nor too slow?
- ❑ Did I speak clearly and audibly?
- ❑ Did I vary the tone of my voice to make the presentation interesting?
- ❑ Did I make appropriate eye contact with the audience?
- ❑ Did I complete the presentation within the time limit?
- ❑ Did I answer questions fully and succinctly?

13.5 Step 4: delivering your presentation

13.5.1 Dressing appropriately

Make sure that what you are wearing is suitable for the presentation. For example, if it is a formal presentation such as a scientific meeting, then you will need to dress formally. If it is a departmental seminar, then informal wear may be appropriate.

13.5.2 Managing your nerves

Most speakers, no matter how experienced, feel some degree of nerves just before they speak in public. However, if you prepare carefully and rehearse your presentation until you are able to speak fluently, this will reduce the anxiety. In addition, you can use the following techniques to manage your nerves.

- **Breathe deeply**. A simple breathing technique which can be used to help you relax is to breathe in deeply and then let your breath out. If you do this 10–15 times before the presentation, it will help reduce any nervous tension.
- **Visualize delivering a successful presentation**. You can use imagery to help you prepare for the presentation. This works by visualizing yourself going through the entire presentation in your mind from start to finish successfully. Going through the presentation mentally before the actual event will help you feel more prepared and self-assured about delivering the presentation on the day.
- **Drink water**. It is common to feel dry-throated before a presentation as a consequence of increased adrenaline. If you keep a glass of water near you and sip before and during the presentation (if you need to) then this will assist in alleviating the dry throat.

13.5.3 Using body language carefully

Your body language—that is, the way you stand, your facial expressions, and whether you make eye contact with your audience or not—contributes to how the material of your talk is received by the audience. The following are some tips for using body language to communicate your material confidently and to establish a rapport between the speaker and the audience.

- **Body posture**. When delivering your presentation stand tall with your shoulders back, head up, and arms at your sides. By standing in this manner you will project an air of confidence (even if you are not feeling it). Do not slouch, look down, or stand with folded arms as this stance will portray to the audience that you are nervous. This will not instill confidence in the audience about you or what you have to say. Avoid any distracting mannerisms, such as rocking on your heels.
- **Stand in the middle of the floor if at all possible**. If you are standing at a lectern then stand away from it unless you are using a fixed microphone. Always face your audience. If you need to point to the screen then turn, point, and then face the audience again. Avoid talking with your back to the audience.
- **Make eye contact with your audience and smile**. This is extremely important in establishing a rapport with the people you are speaking to. If you are speaking in a large room, then aim to look at the back of the room and sweep the audience from right to left so that you face the entire audience. Never stare at any one person or face only a part of an audience. Aim to look at each part of the audience for at least 2 seconds before moving on to the next part (this is roughly the length of time it takes to communicate one complete idea to the audience). Remember, it is more difficult for people to look out of the window if they think you will notice!
- **If you are using a pointer do not wave it around**. Switch it off when you are not using it. If you have an unsteady hand, then you can make it more steady by holding with your free hand the elbow of the arm which is holding the pointer. Remember, if it is a laser pointer, never point it at the audience as this is dangerous.

13.5.4 Using your voice carefully

When speaking, you should project an energetic voice that sounds confident and authoritative and which will engage the interest of your audience. The following is a list of tips for using your voice to complement your presentation material and your body language. For further guidance on using your voice appropriately, see Payne (2004).

- **Speak loud enough so that all in the room can hear you.** Always try and project your voice to the back of the room. If the room is large and you are not able to project your voice, then use a microphone. However, even with a microphone, you will need to project your voice outwards. If you are not familiar with the use of microphones, then practice using one before the actual delivery of your presentation.

- **Speak clearly** so that the audience can identify each word that you speak. Use simple language as far as possible. Do not mutter and avoid terms such as 'okay', 'uh', 'sort of', and 'you know'.
- **Pace your delivery** so that you neither speak too quickly and rush your delivery nor speak so slowly that the audience loses interest in what you are saying. Speaking too quickly can be a sign of nerves and you may therefore want to use some of the techniques for managing this as described earlier. Speaking too slowly can be a sign of insufficient practice, and hence practising until you are able to speak fluently can speed up the rate at which you speak. It is always a good idea to get feedback on the speed of your delivery during rehearsals from colleagues, friends, or family and then to modify your pace accordingly.
- **Use variety in your speech**. When you speak, think about how you could vary the volume, tone, and speed of your speech to maintain the interest of your audience. For example, you may emphasize key words or messages by pausing after you have delivered them, or you may increase the volume of your speech when delivering a key fact. Do not speak in a monotonous tone as the audience will lose interest in the talk and will stop listening to you.

13.5.5 Introducing your talk and finishing it clearly and concisely

- **It is always a good idea to memorize your introductory sentences and your finishing sentences.** This will ensure that the start and end of your presentation are clear and confident. If the chair introduces your talk by using its title, do not start the talk by repeating the title—be prepared to modify your beginning when memorizing your introductory lines.
- **Always end the presentation clearly and decisively** so that the audience is sure what the take-home messages are and knows your presentation has finished. A common mistake in presentations is to trail off at the end. A decisive, practised conclusion will ensure that this does not happen.
- **Never read from your notes directly.** Remember, your notes are there to serve as a prompt in case you forget what you were about to say; therefore, you should never read directly from your notes or your slides. Instead, use your slides as an 'aide memoire': for example, if your slide is text based, then talk around each bullet point. If your slide is a figure or table, then explain what the figure is showing. This could be by highlighting the key trends of the data or, if it is a schematic, then talking through the diagram. Try to be relaxed—remember the audience is there because they are genuinely interested in your work. Therefore, make your delivery a pleasant and informative experience for them.

13.6 Step 5: answering the questions

Most formal presentations are followed by a question and answer session. This session provides the audience with an opportunity to ask for clarification, to make comments, or to criticize your work. It gives you the opportunity to demonstrate your understanding and to explain and defend your work.

If the session is informal, then you may be able to take questions throughout the presentation. However, some people find this distracting and it can break up the flow of their presentation. You should let the audience know before you start your actual presentation whether you prefer if questions are left till the end or if you will take questions throughout the presentation.

The following guidelines will assist you in managing the questions and answers session effectively.

- Try to anticipate the questions you may be asked before the presentation and think about what your responses to these questions could be.
- Listen to the question being asked carefully. If you do not hear a question fully, then ask the questioner to repeat it. If you do not understand the question, ask for clarification.
- Wait for the questioner to finish asking the question before you start to answer.
- Answer the question clearly and succinctly. Do not be vague and ambiguous in your response.
- If you don't know the answer to a particular question, then say so. You could follow this up by offering to get back to the questioner after you have researched the answer. It is always better to admit you don't know something than to make up an answer—an answer which at least some of your audience may be able to tell is incorrect!
- Always be respectful to the questioner (and to the audience). Make eye contact with the questioner when you are answering their question, but do not forget to focus on the remaining audience as well.
- Do not become defensive if someone criticizes your work. See it as an opportunity to explain your work, or for you to think about an issue you may not have considered. Questions can frequently make you think about your work in a new way.

13.7 Step 6: evaluating your presentation

If you are to improve your oral presentation skills then it is necessary to reflect on how successful your presentation was. One way of doing this could be to ask for feedback from any friends and colleagues who attended the session. You could also use the review checklist in Box 13.3 to self-evaluate your performance. By reflecting on the effectiveness of your presentation, you will be able to identify areas you need to focus on to improve the quality of any subsequent oral presentations.

References

Fraser, J. and Cave, R. (2004). *Presenting in biomedicine: 500 tips for success*. Radcliffe Medical Press, Oxford.

Payne, R. (2004). *Vocal skills pocketbook*. Management Pocketbook, Alresford, Hants.

Walters, D.E. and Walters, G.C. (2002). *Scientists must speak: bringing presentations to life*. Routledge, London.

Chapter 14

Preparing and presenting a research poster

Introduction

The aim of this chapter is to guide you through the process of presenting your research findings in the form of a poster display.

Specifically, this chapter will:

- describe the content and style of a poster
- provide guidelines on presenting the poster at a scientific conference or symposium
- include an example of a well-designed poster.

14.1 Poster presentations: an overview

The objective of a research poster is to communicate your research findings in a visual format. Poster sessions form an integral part of scientific conferences (see Chapter 13, Table 13.1) and are used by the researchers to communicate initial research findings which have not yet been published as a full research paper. If you want to present a poster at a conference, then you will need to submit an abstract to the conference organizers by the abstract submission deadline (see Chapter 10, section 10.3). If the abstract is selected, then it will be published in an abstract book and you will be able to present your findings as a poster display. Poster sessions are not limited to conferences but also form part of graduate and undergraduate departmental symposia. Again the aim is to communicate your research findings visually to other colleagues attending the symposium. Presenting posters at departmental symposia is an excellent way of disseminating research findings in a small and generally supportive environment. It will develop your poster presentation skills and your ability to answer critical questions about your work. Both skills will prepare you for presentation at a conference where the audience could be bigger and much more diverse.

At a typical poster session, posters from different authors are exhibited in a poster room. The posters are mounted and displayed before the poster viewing session.

During the actual viewing session, the authors stand next to their poster while the audience circulates around the room reading the posters and asking the authors questions about their work.

The poster viewing session (and indeed the conference or symposium as a whole) serves a number of valuable purposes. It provides you with the opportunity to communicate your research findings and discuss them with the colleagues attending the conference or symposium. In addition, it provides you with a snapshot of the most current research in the field and the opportunity to meet and speak to other researches who are working in the same field as you or a related one. These discussions can help broaden your understanding and develop new perspectives about your own work, as well as providing you with an opportunity to network (Chapter 15).

The features of a well-designed research poster are:

- It provides a brief overview of the research findings.
- The content is self-explanatory so that the central message of the work can be understood from the display itself.
- The text and graphics are clearly legible to a reader standing approximately 1–2 metres away from the poster.
- The design is simple and clear but at the same time attracts the attention of the audience.

There are four steps to designing and presenting a good poster, and these are presented as a flow chart in Figure 14.1, starting with the planning stage of the poster and then moving on to preparing the poster. This is followed by presenting the poster at the poster

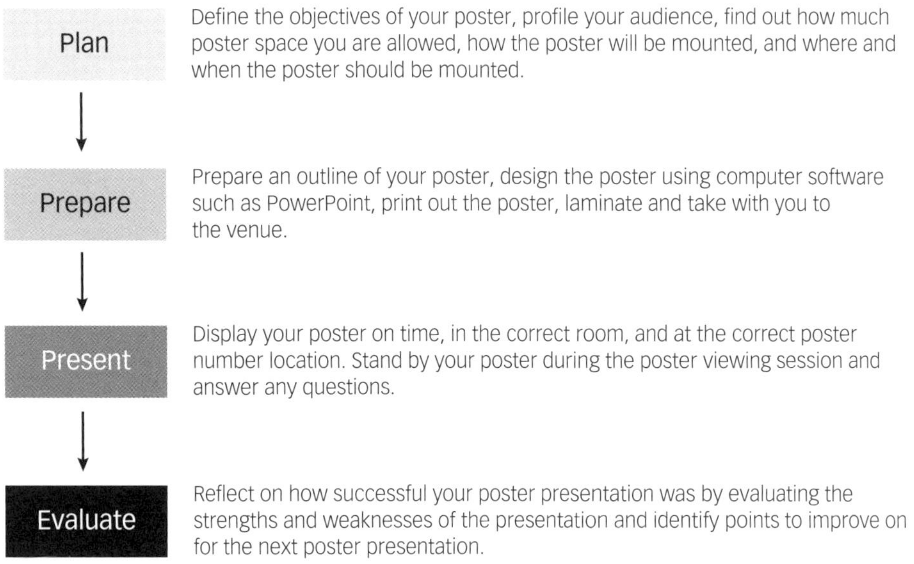

FIGURE 14.1 The four stages of designing and presenting a poster.

session and finally evaluating how successful the poster presentation was. The following sections of this chapter describe each of these stages in turn.

14.2 Step 1: planning the poster

The first step to designing an effective poster is to plan. When planning, address the following four questions:

1. **What** are the objectives of my poster?
2. **Who** is the audience?
3. **How much** poster space is allowed and how is the poster mounted on the display board?
4. **Where and when** should the poster be displayed?

14.2.1 What are the objectives of my poster?

The objective of a scientific poster is to communicate your research findings in a visual format. Therefore, aim to convey the central message of your work in a simple and clear way but at the same time make the poster visually appealing.

14.2.2 Who is the audience?

The composition of the audience will depend on whether the poster is to be presented at a conference or a departmental seminar or symposium. If you are presenting at a conference, it is likely to be an expert audience with a knowledge base in the topic of the meeting. At a departmental symposium, the audience is likely to be mixed, composed of experts as well as people who are working in the same discipline but not necessarily the same field as you. If your audience is mixed, then include some general and some specific points in the introduction to your poster to make it more accessible to the audience.

14.2.3 How much poster space is allowed and how is it mounted on the display board?

Poster sizes can vary, and the exact dimensions will be stipulated by the conference or symposium organizers. Examples of sizes are 90 cm × 90 cm, 91cm × 152 cm, or 120 cm × 180 cm, and these can be in landscape or portrait orientation. Posters are mounted on a poster display board using mounting material such as Velcro or drawing pins.

Before you start preparing your poster, find out how much poster space you are allowed and if your poster should be in a particular orientation. Make sure the dimensions and orientation of your finished poster conform exactly to the instructions supplied, otherwise you may find that it does not fit on the poster display board. If you are given the dimensions of the poster display board instead of the actual poster size, make the poster slightly smaller than the size of the board itself. In addition, find out how the poster will be mounted (e.g. Velcro or drawing pins) and whether these will be supplied at the venue or not. In any case it is best to take the mounting material with you as back-up, just in case.

14.2.4 Where and when should the poster be displayed?

You will be given a venue for displaying the poster and a time by which the poster should be mounted on the display board. You will also be allocated a poster number which is the exact location in the poster room where your poster should be mounted. Make sure you mount your poster at the stipulated location and by the time you have been given. The location will correspond to the number allocated to your abstract in the published abstract book (see Chapter 10, section 10.3) which everyone attending the conference will be supplied with. This ensures that people who are interested in finding out more about your work can find you and your poster.

14.3 Step 2: preparing the poster

Once the planning is complete then you can move on to preparing the poster itself. There are three parts to preparing your poster:

1. Prepare an outline of your poster.
2. Design the poster using computer software such as Microsoft PowerPoint.
3. Put it all together, ready for transporting to the meeting.

14.3.1 Prepare an outline of your poster

First decide on the content of your poster taking into account the size and typical components of a poster. Similar to a research paper, a research poster follows a standard structure composed of a title, abstract, introduction, methods, results, conclusion, references, and acknowledgements. A brief description of each of these components is given in Table 14.1.

To help you decide on the most appropriate content for your poster, ask yourself the following question, and use the content of your pre-submitted abstract to guide you:

- What is the central message I want to get across to the audience? Try to aim for one central message (this will be your key conclusion).

TABLE 14.1 Components of a poster reporting research findings

Title (author and affiliation)	The title of the research work, the people involved in the work, and their places of work should be placed as a banner heading that spans the full width of the poster. An institutional logo can also be placed in the title banner
Abstract	This should clearly summarize the content of your poster and include the aims of the research, the main methods, the main results, and the main conclusions (Chapter 10). Note: not all posters include an abstract—check the instructions supplied by the conference organizers to see whether one is required
Introduction and research objectives	Identify clearly the objectives of your research and provide sufficient background to the study to place your work in context. Make sure you clearly establish which gap in knowledge you are addressing and why it is important to address it
Methods	Describe the experiments performed to investigate the problem. They should be presented briefly and in a format that can be easily understood by the audience, e.g. in diagrammatic form
Results	Present the key data in illustrative format (figures or tables). Use a maximum of six figures or tables and add descriptive titles and legends
Conclusions/summary	Add a short conclusion or summary of the work at the end to identify your main findings. You may also make suggestions for further work (optional)
References	Include a short list of references as necessary
Acknowledgements	Add a very brief statement of acknowledgement identifying the organizations that funded the work and people who provided any additional support such as technical assistance. Institutional logos can also be used at this stage

- Which results are essential to support my conclusions?
- Which methods are essential to understanding my results?
- What gap in knowledge is this study addressing?
- What is known at the end of the study that was not known before starting the study?

Once the material is selected, then consider how to lay out the material. An example of a typical layout is shown in Figure 14.2. In this example, the information is arranged in landscape orientation. The title banner spans the full width of the poster and four panels of information are placed below it. Each panel is read from left to right and top to bottom. If you are using a layout which should not be read from top left to bottom right, then

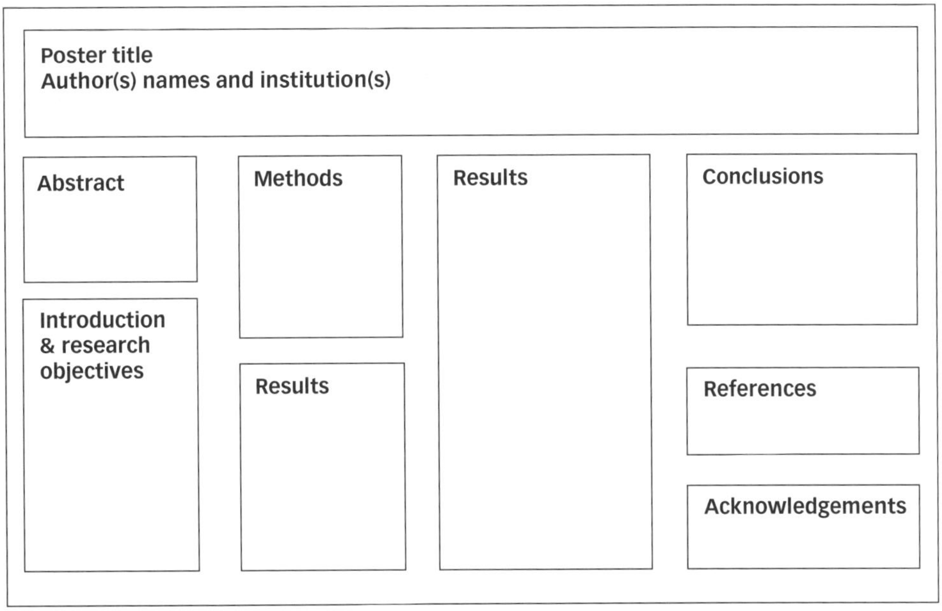

FIGURE 14.2 A typical layout for the components of a research poster.

number each box in order or put arrows between the boxes to help the reader navigate the poster. You could also consider substituting the traditional headings (Introduction, Methods, Results, and Conclusions) with information statements. For example, instead of entitling a section 'Methods', you could use a heading such as '*We studied two cohorts of lambs in Northern England in 2006*'. This will signpost the different sections of your poster and make it easier to follow.

14.3.2 **Design the poster**

Once you have decided on the content and the arrangement, then move onto designing the actual poster itself. A poster can be designed in two formats:

- A single large sheet format.
- A split format consisting of a number of single sheets.

The single large sheet format is now more commonly used and can be designed using software such as Microsoft PowerPoint. If you are unfamiliar with using PowerPoint to produce a large-format poster then consult the textbook by Irwin and Terberg (2004) for a series of simple step-by-step instructions. If you design the poster as a series of single sheets, you can still use Microsoft PowerPoint to produce your text and graphics.

When designing your poster, aim for a simple design and use a combination of text and illustrations to make it visually appealing to the audience. Tips for preparing a good poster are outlined in Box 14.1.

BOX 14.1 Good poster design—some tips

Writing the text

- Titles should be legible from 5–10 m away. For example, use a font size of 96pt for the main title of the poster and 72pt for the names of authors and their place of work.
- Use a font size of 48–54pt when writing sub-headings such as Abstract, Introduction, Methods, Results, Conclusions, References, and Acknowledgements.
- For other text (including figure legends) use a font size that is legible from a distance of 1.5–2 m (24–28pt).
- Use bold type for the main title and the sub-headings.
- Use a sans serif font such as Arial, Tahoma, or Verdana as these are the most legible.
- Use a single font (or at most two) throughout the poster, including the title, sub-titles, and figure legends.
- Write the main text in single spacing.
- Keep the text brief and concise. Use bullet points if this adds clarity to the poster.

Using colour

- Aim for a high contrast between the background and the text. For example, use dark text (black or dark blue) on a light background (e.g. pale blue, pale yellow, or white).
- Use a single background colour without any patterns or images.
- Use a limited number of colours throughout your poster (three or four colours is more than sufficient). When selecting colours opt for those that harmonize with each other and avoid those that clash. Bold colours such as red, green, and blue are best.

Preparing figures and tables

Chapter 11 provides guidance on preparing accurate and comprehensive figures and tables. However, bear in mind that figures and tables prepared for display in a poster must be much simpler than those prepared for publication in a paper, so that the trends or items can be visualized quickly and easily. Some key points to consider when preparing tables and figures for poster display are:

Tables

- Keep the table small and simple. For example, The Physiological Society guidelines for designing posters state that no more than 30 values should be included in a table.
- Label each table column with an informative heading and include units of measurement (as necessary).
- Include a brief title and footnotes to make the table self-explanatory.

Graphs

- Use no more than three lines of data on a single line graph.
- Use bold lines and contrasting colours for each data line in a line graph. Avoid using fine dotted or dashed lines as these are difficult to read.
- Use the same colours for the same treatments in different graphs.
- Label each axis, include units of measurement, and add standard errors (if necessary).
- Include a key to explain the different data sets.

Micrographs, maps, electrophoretic gels, and blots

- Crop the image to select only the relevant information.
- Add explanatory labels to each lane in a gel or blot.
- Label each significant band in a gel or blot with arrows, name, and/or size (as appropriate).
- Mark micrographs and maps with arrows to highlight any significant features to direct the attention of the viewer.

Line illustrations

- Make line diagrams simple and clear to read.
- Aim for bold and clear lines.

Include brief and informative legends to make the figures self-explanatory.

Arrangement of poster components and space

- Arrange the sections of the poster so that they can be read smoothly. The easiest to read is from left to right and top to bottom, so arrange sections to read in this direction (Figure 14.2). If you use a different layout, then number each section in order or put arrows between them to help the reader navigate the poster.
- Aim for a poster that is made up of approximately 20% text, 40% graphics, and 40% empty space. Do not overload your poster with too much information. It will end up looking cluttered and messy, and will be difficult to read.

14.3.3 Putting your final poster together

Once you have produced your initial draft, then review and redraft your poster, using the checklist provided in Box 14.2. A common mistake in poster design is to include too much text. Therefore, go through your poster carefully with the following points in mind:

- Edit out any non-essential words and make the poster as concise and as informative as possible.

BOX 14.2 Reviewing your poster—checklist

Content

- ❑ Does the poster include the essential components: title, introduction, methods, results, conclusions, references, and acknowledgements?
- ❑ Is the title concise and informative?
- ❑ Are the objectives of the research clearly defined at the beginning?
- ❑ Is the central take-home message clear?
- ❑ Are all the tables, graphs, and line drawings appropriate and relevant?
- ❑ Is your poster free from spelling mistakes, and grammatical and punctuation errors?
- ❑ Have you checked your figures and tables for completeness?
- ❑ Are graph axes labelled?

  Do all table columns have headings?

  Are units of measurement included?

  Are standard deviations or standard errors included where required?

  Is there a key explaining any symbols, shadings, or line colours you have used?

  Are any bands and lanes clearly labelled?

  Are any significant features on the images clearly marked?).

Design

- ❑ Is the text legible from a distance of 1.5–2 m for the main text and 5–10 m for the title?
- ❑ Is the font easy to read?
- ❑ Is there a good balance of text, graphics, and space on the poster?
- ❑ Are colours used effectively so that the poster looks coordinated and professional?
- ❑ Are the components of the poster arranged so that they can be read smoothly?

- Consider whether any text can be replaced with a figure as these are usually easier to follow and take up less space.
- Consider whether any continuous paragraphs of text can be rewritten as bullet points.
- Consider substituting headings such as Methods, Results, and Conclusions with informative statements. For example, instead of 'Results', consider a statement such as 'Wheat grown in high nitrogen supply grew faster than . . .'.

Once the design of the poster is complete, you are ready to print it out. If you are creating your poster as a single large sheet, then you will require access to a large-format printer. Your university is likely to have suitable printing facilities. The printed poster can be laminated for durability and carried rolled up in a poster tube.

If you are creating your poster as a series of sheets, print out each sheet as A3 or A4 either on card or on paper. You can then cut each sheet to the desired size using a paper cutter. Again, laminate each individual sheet to increase durability.

14.4 Step 3: at the poster presentation session

You must mount and display your poster on time, in the correct room, and at the correct poster location. During the session, stand next to your poster to answer any questions that may be addressed to you. To make the most of the poster presentation session, you could:

- Prepare a brief talk (e.g. 3–4 minutes) on the subject of your poster. Some conference attendees prefer to hear you talk them through the poster rather than having to read it themselves.
- Try to anticipate the questions you may be asked before the presentation and think about what your responses to these questions could be. Types of questions could be: Why did you do this study? How did you conduct this particular experiment? What does this statistical analysis mean?
- Have handouts of your poster ready for anyone who may be interested in your work. If your poster is in single sheet format, then you can simply reduce it and print it out as an A4 sheet (but make sure the text and graphics are still legible at the reduced size). If your poster is constructed as a series of sheets, or cannot be reduced to A4, then you could print it out as a series of PowerPoint slides.
- Include your name and e-mail address on the handouts to encourage further contact. Alternatively (or additionally) have business cards ready to hand out with your contact details for anyone who may decide to follow up your work with you after the session.
- View the other posters in the room and take down any contact details of people whose work you are particularly interested in.

14.5 Step 4: after the poster presentation

If you are to improve your poster presentation skills then it is necessary to reflect on how successful your poster presentation was. One way is to identify two strengths and two weaknesses of your poster presentation. This way you identify those areas which are strong, and therefore you should continue with, and those areas you need to improve on the next time you prepare and present a poster. It is also a good idea to look at other posters displayed at the poster session to pick up any additional good points that could be incorporated into your own poster design.

After the poster presentation, if you have promised a conference attendee to provide them with further information, then follow up on this. Also follow up with any presenters whose work you are interested it. This could lead to further discussions and even productive collaborations (Chapter 15).

14.6 Example of a well-designed poster

Figure 14.3 shows an example of a well-designed research poster. It was presented at the World Association for the Advancement of Veterinary Parasitology Congress, Gent, Belgium, 2007.

For full-colour version of the poster, refer to Online Resource Centre

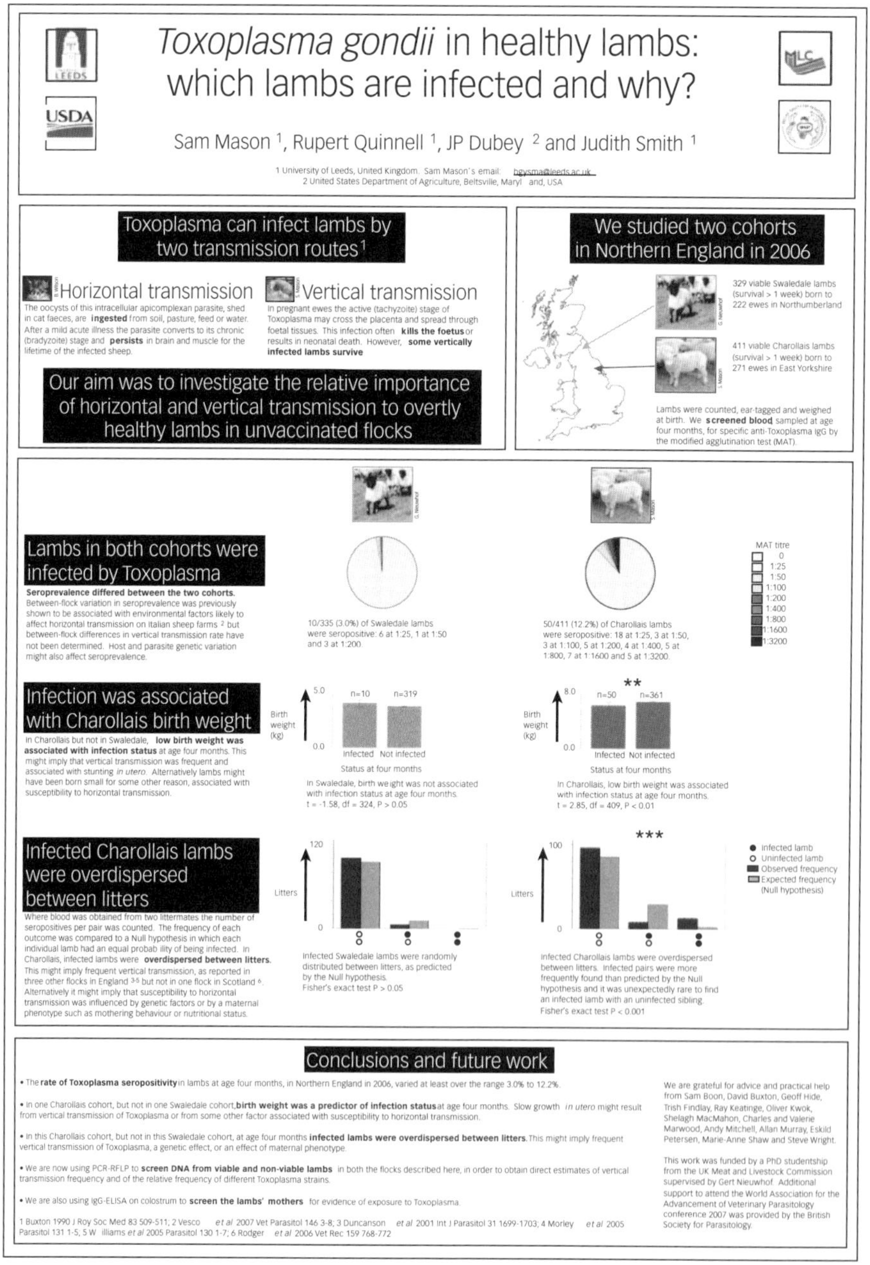

FIGURE 14.3 An example of a good poster.

If you read through the poster and review the content and design, you will see that the poster is well designed because:

- It is simply and clearly presented but at the same time it is visually appealing.
- There is a good balance of text, graphics, and space.
- The font and type size make it easy to read, and colours are used effectively so that the overall effect is a well-coordinated poster which looks professional.
- It includes the essential components: title, introduction, results, methods, conclusions, references, and acknowledgements, arranged in a way which makes the poster easy to navigate.
- Traditional headings such as Introduction, Methods, and Results are not used but the poster is clearly signposted using informative statements instead.
- The objectives of the research are clearly defined and the central take-home messages are clear from the poster.

If you use the guidelines presented in this chapter, you should similarly be able to produce a high quality and professional poster (and could even win a prize for best poster!).

Reference

Irwin, T. and Terberg, J. (2004). *Perfect medical presentations: creating effective PowerPoint presentations for the healthcare professional.* Elsevier, London.

Chapter 15

Networking

Introduction

This chapter outlines the benefits of networking and describes some of the ways in which you can network.

Specifically, this chapter will:

- explain what networking is and why it is important to network
- outline the different ways in which you can network, including face-to-face networking and the use of web-based technologies
- describe some web-based tools for collaborative authoring
- describe how to network effectively and ethically online.

15.1 What is networking?

Networking is an important part of your academic career. It involves developing and maintaining contact with people who work in the same field as you or in a related field. The reasons for networking are:

- **It expands the number of people with whom you can share information and ideas informally**. These exchanges can be very useful in terms of enhancing your understanding of your work and the work of others. In addition, it could lead to a new job or a meaningful collaboration with other researchers. The types of information you may share are:
  - technical expertise and reagents
  - references
  - research data and new research ideas
  - tips on teaching
  - publishing and funding opportunities

    - recruitment opportunities
    - forthcoming events
    - opinions on new developments and topical scientific issues in your field of work.
- **It makes you and your work more widely known**. Reporting your results to the wider scientific community is the key way by which you build your reputation as a scientist. During the course of your research, you will disseminate your results initially through oral and poster presentations at scientific meetings and then through written papers that are published in peer-reviewed journals (see Chapter 3, section 3.3). This will make you and your work more widely known and improve your chances of securing employment (and research funding at later stages of your career) (Boss and Eckert, 2003).
- **It provides you with a support network** so that you have like-minded people with whom you can share research and study experiences.
- **It is good fun** and can lead to productive relationships which can continue after your programme of study has come to an end.

You should start networking at the early stages of your research programme and continue networking throughout your career. There are a number of ways in which you can do this, ranging from informal discussions with colleagues during coffee breaks, through to attendance at conferences and publishing your work through formal channels such as conference abstracts and research papers. An additional and relatively new method of networking is through the electronic medium using web-based technologies such as discussion forums and blogs. Each of these methods, and its purpose and benefits, is described below.

15.2 Networking opportunities

A most useful and immediate networking opportunity is the tea room or other social spaces where colleagues gather for a break during working hours. This is an ideal way to meet researchers from different research groups within your department and find out what they are working on. This will help you to identify how your work fits in to a wider context and you will learn who to approach for advice and information when you require it. You will find that most researchers within the same department (and even within the same institution) generally cooperate with each other in sharing information such as technical expertise and protocols (see Chapter 1, section 1.4.2). It is important that you know who is working with which methods and in which areas so that you can approach the right person for advice if you require assistance. As you establish your reputation as a competent and reliable scientist, similarly, others will approach you for advice. Of course, there will be some researchers who are less inclined to share information—perhaps because of a clash of personalities or if one member of the group is highly competitive. In these instances you will have to gauge who you can ask advice from. Remember, if someone provides you with technical assistance or contributes to your work in some other way, then you will need to consider carefully whether the contribution merits an acknowledgment or even authorship in any resulting publication (see Chapter 3, section 3.5).

15.2.1 Attending and contributing to departmental seminars

Most university departments or research institutes hold weekly seminars that cover a broad range of topics in the discipline. At these seminars, graduate research students and other invited speakers report their research findings orally. Each seminar is typically an hour long, which includes time for questions and answers at the end of the presentation. By attending the seminars you will broaden your knowledge of the discipline area you are working in and find out who is working in a similar field to you or in a related field. Similarly, if you present your work at seminars, others will find out about it.

Departmental seminars are an excellent way of disseminating your research findings in a small and generally supportive environment. They enable you to develop your presentation skills and your ability to answer critical questions about your work. Both skills will prepare you for presenting your work at a scientific conference where the audience could be bigger and more internationally diverse.

15.2.2 Attending scientific conferences

Scientific conferences (also known as scientific meetings) are usually sponsored by subject-specific societies and attended by researchers working in the same field or related fields. The meetings may be international, national, or local, and will consist of an expert audience with a knowledge base in the topic of the meeting. At these conferences, researchers present their most current research findings, which are not yet published as complete research papers, as either poster or oral presentations (Chapter 13 and 14).

Attendance at conferences will give you the opportunity to present your preliminary data in either poster or oral format and expose you to the most current thinking in the field. In addition, you will have the opportunity to meet and engage in discussion with other scientists working in the same field as you or a similar one. It is therefore extremely important that you attend as many conferences as you can during your research programme. You may decide initially to attend a local or national conference and then, as your confidence builds up, attend an international conference. You supervisor will be able to advise you on the most suitable conference for presenting your data. Conferences are usually sponsored by subject-specific societies, so you can browse through the society websites to find out when and where conferences are taking place. They are also advertised on subject-specific journal websites and usually (though not always) listed on the academic worldwide conference database (http://www.conferencealerts.com/).

Getting the most out of a conference

- Present your work either as a poster or as an oral presentation. If you want to contribute to a conference, then you will need to submit an abstract of your work for consideration by the conference organizers. If your work is selected for presentation, then the abstract will be published as part of an abstract book (see Chapter 10, section 10.3) and can be listed on your CV. In addition, your presentation will generate a productive discussion about your work which can be very useful in helping you develop new insights.
- Talk to other people at the conference. If you are particularly interested in someone's work, then find out more about what they are doing. It could lead to fresh perspectives on your work and even a productive collaboration.
- Take down the contact details of anyone whose work you may decide to follow up after the presentation. Similarly, have business cards ready to hand out with your contact details for anyone who may decide to follow up your work with you after the session.
- If you have promised to provide someone with further information, then you must follow up on this after the session. In addition, you should follow up the initial contact with anyone whose work you are interested in.

15.2.3 Joining a professional scientific society

Scientific societies promote the understanding of their subject by running a range of activities such as organizing conferences, publishing or sponsoring journals in their field, and running workshops. Scientific societies can be general, such as the British Association for the Advancement of Science, or more specialized, such as the Genetics Society or the Biochemical Society. There are many benefits to joining a professional society, for example:

- You will receive updates on developments in your field.
- You will be informed of conference dates and venues.
- You will be able to apply for travel grants to attend conferences and workshops.

You should join a society that is most relevant to your subject area. You can find lists of relevant professional societies by searching gateways such as BUBL (see Chapter 5, section 5.2.3) or, more easily, by asking your supervisor or colleagues.

15.2.4 Attending local networking and discussion meetings

An increasing number of scientists are setting up local groups based on shared professional interests. These groups organize events such as lectures and discussions with invited speakers from industry or academia followed by an informal lunch or dinner

(Watanabe, 2004). The purpose of these events is to encourage networking and to meet potential recruiters from industry or academia. You can usually find out about local networking groups by talking to colleagues in your department. These events may also be advertised on electronic networking sites such as Nature Network (http://network.nature.com/) (see section 15.4.2). This particular site has a calendar of local events, news, and job information currently for two areas (London and Boston), with plans to extend to other regions as well.

15.3 Effective face-to-face networking

Part of effective networking is to find out who is working in the same field as you and who the potential recruiters are. Reading scientific literature and attending seminars, conferences, and local networking and discussion groups will expose you to people working in the same field and to potential recruiters. Once you have identified the people, then you will need to engage in dialogue with them. Some tips to help you communicate effectively with other people in a face-to-face situation (either on a one-to-one basis or as part of a small group) are:

- **Think about what you hope to achieve** when talking to someone. For example, if you are speaking to another researcher at a conference, then the purpose could be to find out about their work. If you are talking to a potential recruiter, then the purpose could be to find out about job opportunities. Identifying the purpose from the outset means that you can prepare the type of questions you may ask and hence get the most out of the conversation.
- **Think about how you present yourself.** If you want to create a positive impression, then think about the image that your dress, body language, and voice will create. These factors were discussed in Chapter 13 (section 13.5) in the context of delivering an effective oral presentation. In summary:
  - Dress in a way that is appropriate for the type of event you are attending. Formal dress is appropriate for a formal conference but more informal wear is acceptable for a departmental seminar.
  - Always maintain good body posture by standing up or sitting up straight. In addition, maintain good eye contact with the person you are speaking to, and smile. This is extremely important in establishing a rapport with the person.
  - Speak audibly and clearly so that the other person can hear and understand you. Speaking clearly is particularly important if you are speaking to someone who may not be a native English speaker.

Collectively, these suggestions should help create a positive image of you.

- **Use communication techniques** such as listening carefully and actively asking questions. This will show that you are interested in what the other person is saying.

Bomzer (2002) describes an additional communication technique, called synchronization. In synchronization, you match your body language, voice, and facial expressions to mirror that of the other person. All of these techniques can help establish rapport and increase the effectiveness of dialogue.

- **Be clear about the information you can share and information which you cannot share.** When data are ready for sharing and how much data should be shared is discussed in Chapter 3, section 3.3, and you may wish to refresh your memory about this. Some key points from section 3.3 as they apply to online communication are summarized in section 15.5 below.
- **Use relaxation techniques to manage your nerves.** Not everyone can chat easily with people they don't know. If you are anxious about speaking to others, then use the techniques described in Chapter 13, section 13.5 to help you relax. Remember, the more frequently you speak to other people, the easier it will become to initiate and maintain conversations.

15.4 Electronic tools to support networking

An additional way of networking is through electronic media using web-based technologies such as mailing lists, academic networking sites, discussion groups, and blogs. With the exception of mailing lists, the use of these tools to make new contacts and discuss research and scientific issues is not widespread among the scientific community, but it is slowly increasing (Bonetta, 2007). This increase has been facilitated in part by publishers such as the Nature Publishing Group investing in and establishing networking sites such as Nature Network (see section 15.4.2).

There are obvious benefits in using web-based technologies for networking and collaboration purposes. One is that you can establish contact and communicate with large groups of like-minded people who are located in different parts of the world, without leaving your laboratory or fieldwork (Gewin, 2008). Other benefits are that it can save you time, raise your profile in a positive way, provide useful input into your work, and broaden and enhance your perspectives (Declan, 2005; Bonetta, 2007; Crotty, 2008). However, the downside is that you can become bombarded with information and waste rather a lot of time wading through large amounts of irrelevant information. To avoid this, you will need to be selective in the mailing lists you join and the discussion groups and blogs you read and contribute to.

Another concern is the reliability of the material posted on these sites. Some can be highly authoritative (if written by experts), but other material may be factually incorrect. In addition, you should ask yourself how useful to you is the time spent on these activities—would it be more helpful to your career to spend your time in some other way? These and other ethical questions raised by open and collaborative technologies are discussed in section 15.5. Remember, the aim of using these tools is to assist you in networking. They should not replace traditional face-to-face networking, or overwhelm you with information, or distract you from your bench or field research.

The various web-based tools and how they can be used to purposefully and effectively are described below.

15.4.1 Mailing lists

Mailing lists consist of a list of subscribers who exchange views, share information, and announce events through e-mail. Mailing lists focus on a single subject and therefore the discussion between its members is limited to the topic areas within the subject. Academic mailing lists within the UK are listed on the JISCmail website (http://www.jiscmail.ac.uk/) and you will be able to find out about other mailing lists relevant to your work by consulting colleagues. Mailing lists are a useful way to exchange information quickly (for example, to ask questions or make announcements) with like-minded colleagues. The downside is that some mailing lists can be very active and you could receive a large volume of mail daily. Therefore, be selective in the mailing lists you join. Before you subscribe to any mailing list, check the relevance and quality of the content by browsing through the archived postings to see if it is worth joining.

Some simple guidelines for using mailing lists

- Keep the content of the message brief and focused on the subject the mailing list is created to discuss.
- Read the archived e-mails to understand the audience with which you will be communicating. This will help you decide what type of information is appropriate for a particular mailing list.
- Always be polite and respectful to others when writing messages (see netiquette guidelines in section 15.5).
- If you are going to be away for an extended period of time it may be worth suspending a mailing list, particularly if it is very active and generates a lot of mail daily. This will stop your inbox clogging up and you can always ask for archived e-mail messages to be sent to you or view the archived messages from the mailing list homepage when you return.
- Read the guidelines on how to use e-mail effectively to communicate with others (Chapter 2, section 2.2).

15.4.2 Academic networking sites

There are a growing number of networking websites for academics which provide a forum for connecting with people working in the same field as you, sharing resources, and discussing views. These networks are also important for gaining career advice and finding jobs. Examples of useful networking sites are:

- Nature Network (http://network.nature.com/). This site is a product of the Nature Publishing Group and provides a platform for connecting with other scientists through discussion groups, blogs, and profile searches. This site is becoming increasingly popular with young scientists (Smaglik, 2007a).
- PrometeoNetwork (http://prometeonetwork.com/). This is a network for life science researchers and doctors. Members build profiles and become part of a searchable database through which they can discover new colleagues, communicate with their peers, and find new job opportunities. PrometeoNetwork hosts several sub-groups based on shared interest such as 'stem cells' and based on nationality such as 'West Coast life sciences online' (Fais, 2007).
- BioMedExperts (http://www.biomedexperts.com/). This site contains a searchable database of expert profiles of researchers and scientists working in the biomedicine field.
- The blog (15.4.4) *Nascent* (http://blogs.nature.com/wp/nascent/) maintained by the Nature Publishing Group describes new web technology in science, and the blog *SciTechnet* (http://scitechnet.blogspot.com/) maintained by Gerry McKiernan, a librarian at Iowa State University, describes online networking services in science and technology. Both are worth consulting for up-to-date reviews of online networking services.

Creating a profile within a networking site

Most of the academic networking sites listed above will allow you to create a personal profile. The profile typically includes your job description and place of work, a description of your research interests, and a list of your publications. You can add your e-mail address if you would like others browsing through the profiles to contact you. You will also have the option of uploading a photograph of yourself.

The advantage of creating a personal profile within a networking site is that other people browsing the site are more likely to come across you and your work, and this increases the likelihood of like-minded colleagues contacting you.

When creating a profile, be as professional as possible. Keep the content of your profile focused on your work and do not include any personal information. Personalize your web page by adding a photograph of yourself and check the content to make sure it is free from spelling mistakes and grammatical errors.

Add the URL of your profile to your signature on any e-mails you send so that your contacts can view your profile page and find out more about you and your work. You can also add the URL of your profile to your CV so that prospective employers can look through it.

15.4.3 **Online discussion groups**

Discussion groups (also referred to as forums, groups, or newsgroups) are forums for communicating which are created around specific topics. They are similar to mailing lists in that they allow people to exchange views, share information, announce events, and ask questions. However, unlike mailing lists, messages are posted and read from an online 'notice board' instead of the messages being delivered to your e-mail box. This is better as you visit the board when you want and the messages do not clog up your inbox. Discussion groups can be open forums of communication which can be read by anyone, but usually you need to register to become a member and post a message. Some discussion groups are private and therefore visible to invited members only.

You can find a list of groups discussing bioscience research on the BIOSCI bionet (http://www.bio.net) website. The Nature Network website (http://network.nature.com/) also has focused forums and groups discussing various topics in different disciplines. The journal *Science* hosts a careers forum for employment and careers advice.

Joining a discussion group and contributing to the discussions is an effective way of meeting like-minded colleagues, exchanging information, and gaining advice.

Some simple guidelines for using discussion groups

- Before joining a discussion group read through the archived posts to check the quality of the discussions and how relevant they are to your work. Reading the archived posts enables you to understand the community with whom you will be communicating, and this will help you to decide what type of information is appropriate for a particular discussion group.
- Keep in mind that discussion groups are made up of different people discussing different topics with different views. You should therefore be respectful of these views and adhere to the terms of use that govern participation in the discussion group.
- When posting a message keep the content focused on the subject the discussion group is created to discuss.
- Be absolutely clear about what you should not discuss in public forum discussions. This is discussed in section 15.5 under ethical online communication.

15.4.4 **Blogs**

A blog is a website on which information is posted on a regular basis, and typically includes text, images, and links to other web-based information. Blogs can be created and authored by any individual and are used as a method of communicating news and views on current and topical issues in a range of fields. Some people use blogs as personal online diaries in which they document the events of their day. The types of blogs which may be useful to you to read as a bioscientist are:

- **News blogs** which communicate up to date news stories. An example of a news blog is the *Nature* blog (maintained by the Nature Publishing Group) (http://blogs.nature.com/news/thegreatbeyond/) which rounds up and comments on science news from around the world.

- **Subject-specific blogs** which are maintained by academics or researchers on particular subject areas either individually, or as groups of people. To find such blogs browse through some of the following sites:
  - Nature.com blogs (http://www.nature.com/blogs/index.html) maintains blogs on specific topics such as climate change, avian flu, and web technology.
  - Open Wetware Blogs (http://openwetware.org/) hosts a number of blogs on specific and general scientific topics, written primarily by researchers.
  - Research Blogging (http://www.researchblogging.org/) collects blog posts about peer-reviewed research. Postgenomics (http://www.postgenomic.com/) also collects science blog posts and flags up heavily discussed research papers. These are generally written by scientists.
  - Scientific Blogging (http://scientificblogging.com/) and Science Blogs (http://scienceblogs.com/) are websites that cluster science blogs so that they are easier to locate. The writers are mainly scientists and science communicators.
- **Blogs of graduate students and postdoctoral research fellows** who share their perspectives on research and study experiences. They may also highlight important papers, discuss experiments, and comment on broader issues relating to science. Again, browse through the above sites to locate such blogs.
- **Protocol or service-based blogs.** These types of blogs discuss research techniques and protocols, review new products, or provide advice and information on specific subjects. For example, *Bench Marks*, a blog maintained by Cold Spring Harbor Protocols Editor, David Crotty (http://www.cshblogs.org/cshprotocols/) discusses methods used in the biology laboratory. The journal *Science* maintains a *Science Careers Blog* (http://blogs.sciencemag.org/sciencecareers/) which provides advice and information on planning a career in science.

Blogs can play a useful role in bringing you into contact with like-minded colleagues and can provide useful personal insights on the subject you are reading about. They can also provide authoritative opinions on topical scientific issues if the authors are experts on the topic they are writing about. Most blogs allow readers to contribute by posting comments to the blog: this can generate some interesting and productive exchanges. However, the key point to bear in mind when reading blogs is that the material posted there is not peer reviewed and therefore is of variable quality. Some material is written by experts and can be highly authoritative, but other information could be inaccurate or outdated. It is therefore essential that you read the information critically. This should include a consideration of who is writing the blog and their credentials, when the information was last updated, and how objective the information is (see Chapter 5, section 5.2.3).

Writing a blog

Authoring a blog is another method by which you can increase your visibility within the scientific community and communicate your ideas to other scientists and to the general public (Bonetta, 2007). Before you set up a blog consider the following questions carefully:

- **What will you gain from writing a blog?** Benefits of writing a blog are that it will improve your communication skills, you can share comments about peer-reviewed research, and share ideas and perspectives about developments in your research area either with other scientists or with the general public. All of these can broaden and enhance your understanding of science.
- **What will you write about, and who will read it**? This will be determined by why you want to write and who your intended readership is. For example, you may want to discuss a broad range of topical issues for the general public in simple terms or you may want to discuss specific specialist developments with your peers.
- **How much time will you spend on it, and is this time well spent**? Maintaining a good blog can take time, and most people write regularly (either daily or every few days). Writing a good post can take anything from 15 minutes to 1 hour typing time, in addition to thinking time (Yoskovitz, 2007). Therefore, you should consider whether writing a blog will help your career or whether you could be spending this time on other scholarly and research activities. There are examples of how writing a blog can increase your visibility within the scientific community in a positive way. For example, a PhD student was featured in an issue of the *Nature* careers magazine for exploring the best practices of a laboratory website on his blog (Smaglik, 2007b). However, writing a blog can also make you well known for the wrong reasons. For example, a PhD student was threatened with legal action (eventually retracted) for posting copyright-protected figures from a journal on her blog (Gawrylewski, 2007). There are also anecdotal reports of academics being denied promotion or losing out at job interviews as a result of writing a blog (Jaschik, 2005).

Guidelines for writing a blog

- The type of information you write about will depend on the focus of the blog and issues that are important to you. Try to cover a wide range of topics within the focus of your broad theme, and add images and links to your postings to make them more interesting.
- The writing style used in a blog is less formal than that used in a research paper. Avoid heavy use of technical terms, unless you are writing for a specialist audience. This will make your material accessible to a wider audience.
- Write regularly, so that readers know when to expect a post, but do not let your blogging activities keep you away from your research in the field or laboratory.
- Respond to comments posted on your blog. This will ensure a lively and interactive discussion with your readers. It also means that your readers are more likely to return to read your blog.

- Be absolutely clear about what you should not discuss in public forum discussions. This is discussed in section 15.5 under ethical online communication.

Increasing the readership of your blog

- Make the URL of your blog part of your signature/contact details at the end of your e-mail. This will encourage people to whom you send e-mails to read it.
- When posting comments on discussion boards or on other blogs, add the URL of your blog site at the end.
- Add links of other relevant blog posts to your blog. This will ensure that more people link back to you.
- Create a blog within a site such as Nature Network or Scientific Bloggers. These sites are likely to have a larger readership and this will increase the number of people who find and read your blog.

15.4.5 Collaborative web-based tools such as wikis and Google Docs

If you are working on a shared project with other people, then you may want to use web-based collaborative tools which allow you as a group to write and edit the information. There are different types of collaborative tools available such as wikis and Google Docs. Wiki software allows anyone who has editing rights to write, edit, and organize information on a website. A wiki can be open, so that the pages can be viewed and edited by anyone, or restricted, so that the pages are visible but editing is restricted to authorized groups of people. Wiki pages can also be set up so that they are completely private. For examples of how wikis are being used in the biosciences, see the following sites:

- OpenWetWare project (http://openwetware.org/wiki/) is a wiki site where users write courses collaboratively, research groups share and discuss protocols, and interest groups and blogs discuss topical issues. This wiki is open but contributors have to register and establish that they are part of a genuine research organization before they can write any material.
- Scirus Topic Pages (http://topics.scirus.com/) is a wiki site where scientific experts write authoritative material on recent developments on specific topics. This wiki is open-view but editing rights are restricted to authorized people.
- UsefulChem (http://usefulchem.wikispaces.com/) is a laboratory notebook on a wiki site where Bradley's research group at Drexel University record their daily experiments. The wiki is open-view. Bradley's laboratory is conducting Open Notebook Science (Bradley, 2006) and therefore the wiki contains raw day-to-day unpublished experimental results and conclusions (Everts, 2006) (see section 15.5.1).

Another collaborative online tool is Google Docs. This tool allows you to write and share your documents online. It is by default private, but allows you to invite selected people to view or edit a document. You may consider using Google Docs (or similar software) for the following purposes:

- **Writing a thesis.** You can write and edit your work directly online or you can upload edited documents from your computer to your online account. You can grant editing rights to your supervisors, who can then access your files and make comments and corrections. This means you can view changes as they are being made. It also means that you and your supervisor will be working with a single copy of your work. Google Docs documents the changes made, when they were made, and by whom, and you can return to earlier versions of the documents. As the work is stored online, you and your supervisor can access the files from any computer with an Internet connection.
- **Writing a joint paper or proposal.** The functions of Google Docs described above also makes it a useful tool for writing a paper or grant proposal with multiple authors.

15.5 Ethical online communication

When communicating online you must do so in a responsible manner and in a way which does not violate the conditions of your student or employment contract or the conditions of your research funding, or infringe copyright laws. This section provides some guidance on (1) what information you should and should not communicate online, (2) how to communicate with others online, and (3) appropriate use of material that is published on the web.

15.5.1 Communicating with others online

If you choose to engage with the networking and collaborative tools described in section 15.4 then you should be clear about what you hope to achieve and whether this is an effective use of your time. You should use the technologies selectively and responsibly, with full awareness of the potential consequences your words could have. You should also be mindful of any legal restrictions which prohibit you from discussing certain types of information in public. The following are questions to consider.

How useful is it to my career?

Maintaining and contributing to discussion forums, wikis, and blogs takes time and therefore you should consider carefully whether the time spent is helpful to your career or whether your time may be better spent on other activities such as conducting your field or bench research or communicating science through more conventional ways such as writing articles in peer-reviewed journals. Remember, your reputation as a scientist will be measured by conventional activities such as your publication record and presentations at conferences.

How reliable is the information?

If you are using blogs, wikis, and discussion forums to ask questions about aspects of your research (such as advice on how to use a particular protocol), or using information published on sites relating to experimental methods, data, and other information then consider very carefully how reliable this information is. Who is the author? Is s/he knowledgeable about the subject they are writing about or the questions they are answering? Can the accuracy of the information be checked through other information sources? If you do not ask these questions you could end up using inaccurate or outdated information in your work, which could be expensive in terms of time, effort, reputation, and money.

What information can I share and what information can I not share on public Internet forums?

Data sharing was discussed in Chapter 3, section 3.3. The key points are summarized here.

- You can **share your views** on current and topical issues, share your references, provide commentaries on peer-reviewed articles, and discuss methods and products. Ensure that any information you supply is accurate and you attribute any original sources of information which you make use of (see Chapter 3, sections 3.3.3 and 3.4).
- You can also **share your published data** with the online community. This can be a discussion of your work through informal channels such as blogs and wikis. It can also be through depositing your published papers in free-to-access archives and your data sets associated with your published papers in databanks (see Chapter 3, section 3.3.2).
- The vast majority of bioscientists will **not share their unpublished experimental data or their research proposal questions and ideas** (or that of their colleagues) on open Internet forums. The commonly cited reasons are that your experimental findings could be used by another research group who will take the credit by publishing the work before you (Neylon, 2007). Another reason is that material published on these sites has not been authenticated by passing through a peer-review system (see Chapter 4, section 4.2). This means that the information other people are using and building their scientific work on may be of dubious quality. Yet another reason is that some data are so preliminary that they are not at the stage where they can be described or evaluated and therefore are not ready for sharing with the public (Waldrop, 2008).

An exception is Jean Claude Bradley's research group at Drexel University who maintain their laboratory notebooks on an open wiki. This wiki includes day-to-day experimental data including negative and inconclusive results which are shared with the public, and is part of Bradley's Open Notebook Project (Bradley, 2006). Sharing unpublished data on a public Internet forum may not preclude you from publishing your work in some—but not all–peer-reviewed journals, but it will prevent you from patenting it (Everts, 2006; Waldrop, 2008).

- You should not disclose any **confidential or private information** such as that relating to human subjects or proprietary data on public Internet forums.
- **You should not write anything which could harm the reputation** of your institution, your colleagues, or your supervisors. If you have a genuine concern about a particular issue, a public Internet forum is not the appropriate place to raise it. Instead you should use the systems set up in your institution to discuss your concerns.

Overall, be mindful of the ethical and legal restrictions which may restrict you from sharing certain types of information with the online community. Think about what your institutional policies are regarding the ownership of data and the conditions under which your research is funded. If you are in employment, consider your contractual obligations. Remember, disclosing certain types of information could cost you your professional reputation or your job, and could have legal consequences.

Who does the material on the web belong to and can I use it?

Material on the web (websites, blogs, and discussion groups) including text, images, audio, and video recordings is protected by copyright. If you plan to make use of this material, then you should do so in a way that does not violate copyright laws. If you write or develop material on the web, then the copyright may belong to you (as the author) or to your employer or to the publisher. Some copyright owners may make their work available for use under a Creative Commons licence (see Chapter 3, section 3.6, and section 15.5.2).

Terms of use and 'netiquette'

Discussion forums, mailing lists, blogs, and networking sites have terms of use that govern acceptable behaviour which you will be expected to adhere to if you contribute to these sites. Some sites are moderated, which means that if your post or message breaches the terms and conditions of the site it will be removed by the person(s) appointed to act as the moderator for the site.

In addition to the terms of use, there are informal guidelines for communicating on the web. These are termed 'netiquette'. Two simple netiquette rules are:

- Be courteous and respectful when communicating with others online. Do not shout (using upper-case letters is considered shouting) or use abusive or discriminatory language.
- Remember your communication is likely to be read by an international audience; therefore, try not to use colloquial terms or acronyms that cannot be easily understood.

15.5.2 Copyright and the Internet

This section provides some general guidelines on using material from the web in your work and is based on UK copyright law. Before reading this section you should refer back to Chapter 3, section 3.6 on the use of copyright material in your work.

Can I save and print electronic copies of journal articles for research or private use?

Guidelines provided by JISC and the UK Publishers Association (JISC/PA, 1998) indicate that you may download and save a single copy of an electronic journal article from a journal issue, provided it is for your own non-commercial research or private use. You may not, however, download or save a complete journal issue. You may also print out a single hard copy of an electronic journal article from a journal issue. You may save and copy a single chapter from an electronic book, and you may save and print out a single copy of a web page. You may not, however, make further (paper or electronic) copies from that original saved or printed copy

Can I use copyright material in web-based assessments (such as websites and blogs)?

It is acceptable to use copyright material (for example, illustrations or portions of text or sound or performance) in work which is intended for non-commercial assessment. However, the site must not be open to the public as it will not constitute fair dealing if it is. When using copyright material, you must use it appropriately in your work so that you do not lay yourself open to charges of plagiarism. This means discussing the material in the context of your own work and referencing the material appropriately (see Chapter 3, section 3.4).

Can I use copyright material on my personal website or blog (or a discussion forum that I contribute to) which is not for assessment purposes?

Fair dealing allows you to use limited amounts of copyright material for certain non-commercial purposes without requesting permission of the copyright owner and without violating UK copyright laws. The exact amounts are not clearly specified by the Act but some further recommendations are:

- If you want to upload one of your publications on the web, you may do so only if you are the copyright holder or if you are granted permission by the person or body who holds the copyright. If you want to upload articles written by other authors on your website, then you will need to obtain permission before uploading (JISC/PA, 1998).
- If you want to cut and paste some portion of text from another website, then, depending on the amount and purpose of use, you may be able to do so without requesting permission. If the amount is a single extract of 400 words or a series of extracts to a total of 800 words (of which none exceeds 300 words) **and** these are quoted in the context of review or criticism, then this is likely to be considered fair dealing (Society of Authors, 1965). However, it is better to discuss the material in the context of your own work (instead of copying) and hyperlink to the original article.
- If you want to use copyright material beyond that permitted by fair dealing, then check whether the material is covered by a Creative Commons licence (Creative Commons,

2008). As discussed in Chapter 3, section 3.6.3, these licences allow copyright holders to specify conditions under which their work may be used. For example, work covered by the Attribution (by) Licence allows you to distribute, build upon, and display copyright work as long as you acknowledge the original source appropriately. Some Creative Commons licenses will also allow you to adapt the work. Work covered under the Creative Commons licence carries a Creative Commons statement which links through to the conditions of use specified by the author. You should check the conditions of the licence before using the material in your work, particularly if you wish to use it for a commercial purpose. If you want to use the material for purposes and/or in ways other than that permitted by the licence, then you will need to contact the copyright holder to obtain permission.

- Images (such as line drawings, photographs, and maps) are also copyright-protected. You will therefore require permission to upload them. Some images may be covered under the Creative Commons licence and these you will be able to use under the specified conditions. The same applies to sound and video recordings.

Can I use e-mail communication in my work?

E-mail communication is generally private and therefore you should not show or forward e-mails you receive to other people without permission. If you would like to share the content of the e-mail with other people, e.g. in a discussion group or for assessment purposes, then check with the sender to make sure this is acceptable and reference the source appropriately (see Chapter 1, Table 1.2). The same applies if you would like to use the content of the e-mail for publication or for commercial use.

Can I post my dissertation or thesis on a preprint server such as Nature Precedings?

If you post your thesis or dissertation on a preprint server such as Nature Precedings this will not preclude you from publishing the material in some (but not all) peer-reviewed journals. You should therefore always consult the policies of the journal you intend to publish in before posting material on any site. In any event, you should not post on a server (or on a blog or an open wiki) until after your work has been assessed **and** only after agreement from your supervisor(s).

References

Bomzer, D. (2002). How to get a good connection. *Science Careers* [online]. Available at: http://sciencecareers.sciencemag.org/career_development/tools_resources/careers_basics_booklet (last accessed 28 May 2008).

Bonetta, L. (2007). Scientists enter the blogosphere. *Cell*, **129**, 443–445.

Boss, J.M. and Eckert, S.H. (2003). Publishing at the top of the heap. *Science Careers* [online]. Available at: http://sciencecareers.sciencemag.org/career_development/tools_resources/careers_basics_booklet (last accessed 28 May 2008).

Bradley, J. (2006). *Open Notebook Science* [online]. Available at: http://drexel-coas-elearning.blogspot.com/2006/09/open-notebook-science.html (last accessed 11 April 2008).

Creative Commons. (2008). *License your work* [online]. Available at: http://creativecommons.org/license/ (last accessed 7 April 2008).

Crotty, D. (2008). *Web 2.0 for biologists—are any of the current tools worth using?* [online]. Available at: http://www.cshblogs.org/cshprotocols/2008/04/03/web-20-for-biologists-are-any-of-the-current-tools-worth-using/ (last accessed April 2008).

Declan, B. (2005). Joint efforts. *Nature*, **438**, 548–549.

Everts, S. (2006). Open Source Science: on-line communities aim to unite scientists worldwide to find cures for neglected diseases. *Chem. Engin. News*, **84**(30), 34–35.

Fais, F. (2007). Making contacts on-line. *Nature*, **447**, 1140.

Gawrylewski, G. (2007). For blogger: a threat then an apology. *The Scientist*. [online]. Available at: http://www.the-scientist.com/news/home/53177/ (last accessed 11 April 2008).

Gewin, V. (2008). The new networking nexus. *Nature*, **451**, 1024–1025.

Jaschik, S. (2005). Too much information? *Insider Higher Education* [online]. Available at: http://www.insidehighered.com/news/2005/10/11/bloggers (last accessed 7 April 2008).

JISC/PA. (1998). Joint Information Systems Committee and the Publishers Association. *Guidelines for fair dealing in an electronic environment* [online]. Available at: http://www.ukoln.ac.uk/services/elib/papers/pa/fair/intro.html (last accessed 7 April 2008).

Neylon, C. (2007). *Getting scooped. Science in the Open: an openwetware blog on the challenges of open and connected science*. [online]. Available at: http://blog.openwetware.org/scienceintheopen/2007/11/14/getting-scooped/ (last accessed 8 April 2008).

Smaglik, P. (2007a). Nature Jobs. *Nature*, **446**, 825

Smaglik, P. (2007b). Nature Jobs. *Nature*, **447**, 347.

Society of Authors. (1965). [online]. *Quick guide: permissions* available at: http://www.societyofauthors.net/ (last accessed 7 April 2008).

Waldrop, M.M. (2008). Science 2.0: great new tool, or great risk ? *Scientific American* [online]. Available at: http://www.sciam.com/article.cfm?id=science-2-point-0-great-new-tool-or-great-risk (last accessed 7 April 2008).

Yoskovitz, B. (2007). *How much time does it take to blog?* [online]. Available at: http://www.instigatorblog.com/how-much-time-does-it-take-to-blog/2007/03/08/ (last accessed 7 April 2008).

Watanabe, M. (2004). Networking. *Nature*, **430**, 812–813.

Index